Principles of
Atmospheric Science

John E. Frederick
The University of Chicago

JONES AND BARTLETT PUBLISHERS

Sudbury, Massachusetts

BOSTON   TORONTO   LONDON   SINGAPORE

World Headquarters
Jones and Bartlett Publishers
40 Tall Pine Drive
Sudbury, MA 01776
978-443-5000
info@jbpub.com
www.jbpub.com

Jones and Bartlett Publishers Canada
6339 Ormindale Way
Mississauga, Ontario L5V 1J2
Canada

Jones and Bartlett Publishers International
Barb House, Barb Mews
London W6 7PA
United Kingdom

Jones and Bartlett's books and products are available through most bookstores and online booksellers. To contact Jones and Bartlett Publishers directly, call 800-832-0034, fax 978-443-8000, or visit our website www.jbpub.com.

Substantial discounts on bulk quantities of Jones and Bartlett's publications are available to corporations, professional associations, and other qualified organizations. For details and specific discount information, contact the special sales department at Jones and Bartlett via the above contact information or send an email to specialsales@jbpub.com.

Production Credits
Executive Editor, Science: Cathleen Sether
Acquisitions Editor, Science: Shoshanna Grossman
Managing Editor, Science: Dean W. DeChambeau
Associate Editor, Science: Molly Steinbach
Editorial Assistant: Briana Gardell
Production Editor: Anne Spencer
Production Assistant: Sarah Bayle
Senior Marketing Manager: Andrea DeFronzo
Manufacturing Buyer: Therese Connell
Manufacturing and Inventory Coordinator: Amy Bacus
Text Design: Publisher's Design and Production Services, Inc.
Composition: ATLIS Graphics
Cover Design: Kristin E. Ohlin
Cover Image: Courtesy Expedition 13 Crew, International Space Station, NASA
Printing and Binding: Malloy, Inc.
Cover Printing: Malloy, Inc.

Library of Congress Cataloging-in-Publication Data
Frederick, John E., 1949-
    Principles of atmospheric science / John Frederick.
        p. cm.
    Includes bibliographical references and index.
    ISBN-13: 978-0-7637-4089-4
    ISBN-10: 0-7637-4089-6
    1. Atmosphere.   2. Atmospheric physics.   3. Atmospheric chemistry.   4. Meteorology.
I. Title.
    QC861.3.F74 2008
    551.5—dc22

6048

Printed in the United States of America
12 11 10 09   10 9 8 7 6 5 4 3 2

Contents

Preface

The study of the Earth's atmosphere combines a high level of scientific rigor with the ability to observe a complex system on a grand scale. Proceeding from this point of view, this book provides an introduction to atmospheric science for undergraduate students in the physical sciences. The working assumption is that this text is the student's first exposure to the subject. Given the intended audience, the discussion seeks to identify fundamental concepts and principles rather than focus on the finer details of experiment and theory that are focal points for current research. As is the case in a general physics course, the examples given are often simplifications of real-world situations.

The presentation contains a mixture of qualitative and quantitative material. Each chapter begins with qualitative descriptions intended to present underlying concepts. Then, because atmospheric science is inherently quantitative, mathematical derivations place portions of the material on a rigorous basis. The major prerequisite here is knowledge of basic calculus.

Chapter 1 addresses atmospheric composition and structure. There are some fundamental questions here; for example, why should a planet have an atmosphere in the first place? What is the chemical composition of the Earth's atmosphere, and why is it this way? When an atmosphere develops, physical processes rooted in classical mechanics and the kinetic theory of gases create density and temperature distributions with characteristic vertical structures. For example, the temperature profile in the Earth's atmosphere has a well-known dependence on altitude. This profile is a consequence of the properties of gases when acted on by a gravitational field and illuminated by sunlight. This chapter describes the chemical makeup of the atmosphere, the temperature structure, and the processes that govern the vertical distributions of pressure and density.

Chapter 2 considers solar and terrestrial radiation and their interactions with the atmosphere. This material is central to a theory of climate. Since light from the sun supplies essentially all of the energy that drives the Earth's climate system, this chapter begins with a description of sunlight for its own sake and then considers what happens when sunlight interacts with the atmosphere. In addition to receiving energy from the sun, the Earth creates radiation of its own called "terrestrial radiation" or "heat radiation." This radiant energy lies in the far infrared part of the spectrum, where the human eye is unresponsive, but despite being invisible, this radiation is ubiquitous. Summed over a full year, the Earth and atmosphere taken together lose as much energy in the form of terrestrial radiation into outer space as they receive from the sun; this balance between solar energy absorbed and terrestrial energy lost to space is at the core of the theory of climate. An extremely important process called the "greenhouse effect" is an additional issue that, together with the radiation balance, determines the Earth's temperature. The mathematical portion of Chapter 2 derives a simple model to predict the surface temperature of a planet when "greenhouse gases" such as carbon dioxide are present in its atmosphere.

Chapter 3 focuses on water in the Earth-plus-atmosphere system. The Earth is unusual in having large amounts of liquid water on its surface. One of the consequences of this is the presence of a substantial amount of water vapor in the atmosphere, where it has a profound effect on the Earth's climate. Many weather phenomena, including clouds, arise from water changing phase from vapor to liquid or ice. Clouds cover approximately 50% of the Earth, and they play an important role in regulating the planet's temperature by reflecting incoming solar energy back into space. In addition to this reflection, clouds trap upwelling terrestrial radiation that originates at the ground. Water in the vapor phase does this as well, making it an important contributor to the greenhouse effect. Water also has a major influence on the temperature of the atmosphere via its ability to change phase from vapor to liquid or solid. The major derivation in Chapter 3 addresses how these changes in phase influence the temperature structure of the lower atmosphere.

Winds are the subject of Chapter 4, which emphasizes the fundamental scientific principles responsible for the formation of wind systems. Air motions are a response to forces that exist on a rotating planet heated by solar energy. This chapter identifies these forces, describes how they operate, and shows how they combine to create the organized wind systems observed on Earth. Low-pressure systems are particularly important because they are major influences on short-term weather fluctuations experienced at middle latitudes. Day-to-day changes in the prevailing winds and the passage of weather systems are everyday experiences, but a quantitative treatment of atmospheric motions from first principles is a very complicated undertaking. Yet, the fundamental physics is very basic, being Newton's Laws of Motion applied to a rotating spherical system.

Chapter 5 addresses some selected topics in atmospheric chemistry, where the focus is on ozone in the Earth's atmosphere. This chapter describes the chemical theory that explains why an ozone layer exists in the altitude region called the "stratosphere" and how human activity in the second half of the twentieth century led to a loss of some of this ozone. Ozone also forms at the Earth's surface, where it is an important ingredient in the "photochemical smog" experienced by some urban areas. The final portion of Chapter 5 addresses the formation of ozone near the ground.

Finally, Chapter 6 addresses the Earth's climate in broad terms. This chapter provides a definition of climate expressed as the long-term statistics of all quantities required to define the state of the atmosphere, with the timescale to which the averages refer stated explicitly. The early sections summarize the history of climate and climate variability as inferred from the Earth's geologic record. The following sections address mechanisms of climate change, beginning with variability in the incoming solar energy associated with time dependence on the sun and variations in the Earth's orbital parameters. The chapter ends with a brief treatment of the processes that control the atmosphere's carbon dioxide content and of feedback mechanisms in the climate system.

The state of the atmosphere obviously influences the activities of modern society and the quality of life. Examples of these couplings are the demand for energy used in heating and cooling, agricultural productivity, health impact of exposure to gases, particles, and sunlight, and property damage associated with severe weather events. Much of atmospheric science considers processes that influence the human condition on timescales that people can readily grasp and, if necessary, respond to. For this reason, the topics considered in this book can be relevant to policy makers, economists, and corporate leaders as well as to students seeking to understand how physics and chemistry manifest themselves in the atmospheric environment. When one has a grasp of fundamental concepts and principles, it becomes far easier to evaluate the flow of popular information and misinformation to which we are all exposed.

Instructors' Resources to Accompany *Principles of Atmospheric Science*

To assist you in teaching this course and supplying your students with the best in teaching aids, the author and Jones and Bartlett Publishers provide the ancillaries listed below. These ancillaries are available to all adopters of *Principles of Atmospheric Science* and can be downloaded from the Jones and Bartlett website, http://physicalscience.jbpub.com/. The files are located on the text's information page.

- The **Test Bank**, prepared by the author, is available as text files.
- The **PowerPoint® Lecture Outline Slides**, prepared by the author, provide lecture notes, graphs, and images for each chapter of *Principles of Atmospheric Science*. Instructors with the Microsoft PowerPoint software can customize the outlines, art, and order of presentation.
- The **PowerPoint Image Bank** provides the illustrations and tables inserted into PowerPoint slides. With the Microsoft PowerPoint program, you can quickly and easily copy individual slides into your existing lecture slides.
- A list of **Internet Resources** is organized by chapter.

Acknowledgments

Reviewers play a major role in the success of a book. We would like to thank all those reviewers who have spent time commenting on the drafts for *Principles of Atmospheric Science*.

Jon Ahlquist, Florida State University
James P. Boyle, Western Connecticut State University
Will Cantrell, Michigan Technological University
David Flory, Iowa State University
Richard Grimaldi, SUNY, Oneonta
Patrick Market, University of Missouri, Columbia
Gerald North, Texas A&M University
Brian Pettegrew, University of Missouri, Columbia
Steven Sherwood, Yale University
Zhien Wang, University of Wyoming

About the Author

John E. Frederick received his PhD in Astro-Geophysics from the University of Colorado at Boulder in 1975, where he specialized in the physics and chemistry of the Earth's stratosphere and mesosphere. After graduation, he devoted two years to postdoctoral research in the Department of Atmospheric and Oceanic Science at the University of Michigan, Ann Arbor. His postdoctoral years were followed by an appointment as Space Scientist at NASA's Goddard Space Flight Center in Greenbelt, Maryland. During nearly eight years with NASA, Dr. Frederick was active in research related to the stratospheric ozone layer, aspects of chemistry and spectroscopy, and atmospheric remote sensing from Earth-orbiting sensors.

In 1985 Dr. Frederick accepted an appointment as Professor of Atmospheric Science at the University of Chicago. For more than two decades in Chicago, Professor Frederick has guided research by undergraduate and graduate students in areas of radiative transfer, atmospheric chemistry, and atmospheric energetics, with a specialty in studies related to the sun's ultraviolet radiation. He has authored more than one hundred papers published in scientific journals. Professor Frederick's teaching activities include atmospheric radiation and chemistry, environmental science and policy, and general atmospheric science. He is a recipient of the University of Chicago's Quantrell Award for Excellence in Undergraduate Teaching. He is currently Associate Dean of the Physical Sciences Division and of the College of the University of Chicago.

Chapter 1

Chemical Composition and Structure

INTRODUCTION

An atmosphere consists of a mixture of gases and particles subjected to the gravitational pull of a planet and illuminated by radiant energy from a nearby star. Defining characteristics include chemical composition, temperature and its vertical structure, and surface pressure. The observed properties of the Earth's atmosphere arise from the action of sunlight combined with processes taking place near the planet's base, including the oceans, while the presence of life has a profound influence via the production and loss of molecular oxygen. The atmosphere consists of alternating layers in which temperature increases or decreases with altitude. The behavior in each layer arises from the properties of a gas that may be undergoing vertical convection or absorbing energy in the form of sunlight.

The downward force of the planet's gravity acting on a compressible gaseous atmosphere produces a well-defined vertical structure of pressure and density. Air near the ground experiences a pressure created by the atmosphere at all higher altitudes. The associated compression leads to a relatively large density at low altitudes. As altitude increases, the pressure declines exponentially and density shrinks in a similar fashion. For typical air temperatures found on Earth, pressure and density are cut approximately in half for each 5-km increase in altitude.

1.1 Major Gases and Trace Gases

An "atmosphere" is the envelope of gases and particles that surrounds a solid planet, where the force of gravity holds these materials close to the planet's surface. A characteristic length scale that describes the vertical structure of an atmosphere is typically on the order of 10 km (see Section 1.9), as compared to planetary radii measuring several thousand kilometers. Hence, atmospheres constitute thin gaseous layers sitting on top of the planetary surface. This definition of an atmosphere applies to any solid planet as well as to moons in orbit around a planet. It need not refer exclusively to Earth, although the Earth's atmosphere is the focus of this book.

The Earth's atmosphere consists of a mixture of several types of gases. For the moment, assume that the air is completely dry, that is, no water vapor is present. In this special case, the three gases listed in Table 1.1 make up the vast majority of the Earth's atmosphere. For simplicity, Table 1.1 lists the percentage abundances of each of three "major gases" to only one decimal place, although the values are known to still higher accuracy *(U.S. Standard Atmosphere, 1976)*. Molecular nitrogen (N_2) and molecular oxygen (O_2) are by far the most abundant gases, and together these species comprise about 99% of the atmosphere by volume. A third major gas, atomic argon, is produced by radioactive decay of an isotope of potassium in the Earth's crust. Argon plays no significant role in the Earth's atmosphere because it is a noble gas, meaning that it does not participate in chemical reactions. The three constituents in Table 1.1 are well mixed. If one examined a large number of air samples consisting of 1000 atoms and molecules each, on average 781 of these would be N_2 molecules, 209 would be O_2 molecules, and 9 would be argon atoms, for a total of 999. The last molecule would be something else. These proportions are valid from the ground up to altitudes of at least 90 km. No one needs to worry about walking into a blob of pure N_2 and suffocating because there is no O_2 there. Perpetual atmospheric motions keep the various long-lived molecules mixed in the fixed percentages shown in Table 1.1. Above altitudes of about 100 km, however, intense ultraviolet radiation from the sun breaks bonds in O_2. As a result, at sufficiently high altitudes, free atoms of oxygen (O) are more abundant than O_2 molecules.

The chemical composition of the Earth's atmosphere differs from that observed for other planets that orbit the sun. The planets closest to Earth, namely, Mars and Venus, have atmospheres composed almost entirely of carbon dioxide (CO_2). A research topic in planetary science centers on determining why these three planets, formed at about the same time from the same stellar materials, ended up with atmospheres that are so different. Of the trio of rocky planets, Venus, Earth, and Mars, the Earth is clearly the anomaly. Critical ingredients in explaining the unusual chemical composition of the Earth's atmosphere are the presence of large amounts of liquid water on the planetary surface and the existence of life. Section 1.2 considers the role of life in modifying the chemical makeup of the atmosphere, while atmospheric water is the subject of Chapter 3.

TABLE 1.1   Gaseous Composition of Dry Air: Major Gases

Atom or Molecule	% by Volume
Molecular nitrogen (N_2)	78.1
Molecular oxygen (O_2)	20.9
Argon (Ar)	0.9
Total	99.9

Table 1.1 refers to the idealized case of dry air, but the Earth's atmosphere also contains gaseous water (H_2O), in sufficient quantities to qualify as a major gas. Unlike the other major gases, however, the abundance of water vapor near the ground is highly variable with location and time of year. Table 1.2 lists annually averaged water vapor amounts typical of low, middle, and high latitudes derived from values given by Salstein (1995). These averaged numbers mask a large seasonal variability, particularly at middle latitudes. For example, on a day when the temperature is 273 K (32°F), water vapor might make up 0.6% of the ground-level air by volume, while at a temperature of 306 K (90°F), the corresponding value could be up to 4% to 5% by volume. Chapter 3 considers the mechanisms responsible for this sensitive dependence of water vapor abundance on temperature. When water vapor is present in the atmosphere, the percentages of the other three major gases in Table 1.1 must be adjusted downward accordingly since the total always adds up to 100%, including the missing 0.1% that has yet to be identified.

The unique property of water is that it can exist in three different phases, vapor, liquid, or solid, under the temperature and pressure conditions found on Earth, and changes from one phase to another explain the large spatial and temporal variability in the water vapor amount at any location. One of the problems in predicting the magnitude of potential future climate change involves the partitioning of atmospheric water among the vapor, liquid, and solid phases. For example, more water vapor could exist in the atmosphere of a warmer planet, while more water would evaporate from the oceans per year than in the present-day world. The increased atmospheric water vapor abundance would feed back on atmospheric temperatures through processes described in Chapters 2 and 3.

With reference to Table 1.1 for dry air, the sum of the major gas abundances is not quite 100%. To a precision of one decimal place, the total is 99.9%, so something is missing. There are dozens of other types of atoms and molecules in the Earth's atmosphere, and collectively these make up the missing 0.1%. The conventional term for these constituents is "trace gases." A reasonable definition of a trace gas is one whose abundance is less than 0.1% of the atmosphere by volume. In practice, the abundance of any particular trace gas is always considerably less than the 0.1% needed to complete Table 1.1. In fact, the value of 0.1% is an artifact of using only one digit

TABLE 1.2 Typical Water Vapor Abundances near the Ground

Location	% by Volume
Tropics	2.5
Middle latitudes	1.5
High latitudes	0.5

to the right of the decimal point in Table 1.1. With more precise values, the trace gas abundances would actually sum to less than 0.1%.

Some trace gases are created by chemical reactions in the atmosphere, and the field of atmospheric chemistry is devoted to studying these. Perhaps the most important of the chemically active molecules is ozone (O_3), which consists of three oxygen atoms bound together. Near the ground, ozone is present in highly variable amounts. In the countryside, far removed from an urban area, a random sample of 1 billion (1×10^9) air molecules might contain only 20 to 30 molecules of ozone. However, in some urban areas on a summer afternoon, the abundance can be much larger, perhaps in the range of 80 to 120 ozone molecules per billion air molecules for a highly polluted day. Such an ozone amount may appear miniscule, but this quantity can be sufficient to aggravate existing respiratory problems in sensitive people. Chapter 5 considers the chemical processes that lead to these elevated ozone amounts in urban areas.

Most of the ozone in the atmosphere resides at altitudes above $z = 15$ km. From this point onward, the symbol z will refer to altitude above the ground, expressed in kilometers. The chemical mechanisms that create ozone at these altitudes differ from those that operate in urban areas. Near $z = 30$ km, approximately one molecule out of every 125,000 air molecules is ozone. Although this is a much larger relative abundance than exists at the ground, ozone 30 km up is still a trace gas making up only about 8×10^{-4}% of the atmosphere by volume. Ozone at these high altitudes absorbs much of the incoming sunlight in a biologically harmful, ultraviolet part of the spectrum. Still, a fraction of the ultraviolet solar energy survives passage through the ozone layer and reaches the ground. Long-term exposure to this radiation is responsible for a variety of negative effects on human health, such as cataracts and certain types of skin cancer.

Nothing injects ozone into the atmosphere directly; instead, it is formed there by chemical reactions. Other types of trace gases originate at the Earth's surface, where their sources may be biological, geological, or linked to human activities. An important example here is carbon dioxide, the trace gas at the center of concerns about global warming. On the scale of trace gases, CO_2 is very abundant. In 2005, it made up 0.0375% of the atmosphere by volume (Carbon Dioxide Information and Analysis Center, U.S. Department of Energy, http://cdiac.esd.ornl.gov). At these levels, approximately one molecule of every 2650 to 2700 air molecules is CO_2. Furthermore, it is well documented that the amount of atmospheric CO_2 is increasing over time. High-quality measurements of the abundance of CO_2 began in 1957, and at that time it was approximately 0.0315% of the atmosphere by volume. From 1957 to 2005, it rose by approximately 19% to the current value. A well-tuned automobile engine produces CO_2, so the source of this molecule is a natural consequence of the combustion of fossil fuels. Carbon dioxide is a "greenhouse gas," meaning that a growing atmospheric CO_2 abundance should promote an increasing temperature at the ground. The mechanism for this warming is a subject for Chapter 2.

Small abundance does not imply small significance. Atmospheric molecules that are present in very small amounts can nonetheless exert a major influence on conditions at the Earth's surface. Ozone and carbon dioxide are likely more important to the habitability of the Earth than is molecular nitrogen, even though N_2 is more abundant than these trace gases by orders of magnitude.

Most of the material in this book addresses the behavior of gases, but particles in the atmosphere can be significant as well. Historically, interest in atmospheric particles related to their role in the formation of clouds. Particles must be present to initiate the conversion of water vapor to liquid droplets in the atmosphere. More recently, particles have become a major issue in climate change. Particles tend to scatter sunlight back into space and thereby act to cool the planet by preventing absorption of solar energy at the ground. Finally, particles are significant players in some areas of atmospheric chemistry. Certain chemical reactions can occur on the surface of a particle that would not take place if two molecules collided in the gas phase. Although the emphasis of this book is on gases, particulate matter is nonetheless significant in a variety of atmospheric processes.

1.2 Atmospheric Evolution: A Brief Overview

Section 1.1 described the chemical composition of the atmosphere, but why is it this way? This question is a concern of atmospheric evolution, a lengthy topic that this section addresses only briefly. Walker (1977) and Holland (1984) provide comprehensive treatments of the subject from the geological perspective. Important issues here involve the origins of atmospheric N_2, O_2, and gaseous H_2O, the three most abundant gases in the Earth's atmosphere, as well as the trace gas CO_2. Early in the Earth's history, N_2 and H_2O entered the atmosphere as part of volcanic emissions. Modern volcanoes emit both of these gases, and it is reasonable to assume that this was the case with ancient volcanoes as well. Over long periods, a large amount of water vapor outgassed from the interior of the Earth and condensed to liquid to form the oceans. The amount that resides in the atmosphere today is a very small fraction of the total, and there is an ongoing exchange of water among the atmosphere, the oceans, and reservoirs of water located on land, such as polar ice.

Molecular nitrogen is a stable molecule. It is not destroyed efficiently by chemical reactions, and it does not condense to liquid at the temperatures found on Earth. When injected into the early, developing atmosphere, N_2 remained there. Molecular nitrogen is a minor component of volcanic gases. However, over time, it accumulated in the atmosphere, eventually to become the most abundant molecule. This accumulation took place early in the

Earth's history. Today, biological processes lead to an exchange of nitrogen between living things and the atmosphere. Plants utilize nitrogen, which is a major component of manmade fertilizers, although the nitrogen consumed by plants is in a form other than N_2. In the present-day world, various microorganisms convert atmospheric N_2 into other nitrogen-containing molecules that plants can utilize, and because of this biological processing, N_2 has a finite, rather than an infinite, lifetime in the atmosphere. However, any changes in the abundance of atmospheric N_2 occur on time scales much longer than those of interest here.

An examination of planets in orbit around the sun reveals little atmospheric O_2 except on Earth. Something unusual happened here to produce the high abundance of O_2, and this was the evolution of life. A large amount of O_2 exists in the modern atmosphere because of plants performing photosynthesis, a process in which plants make organic matter using sunlight as a source of energy. Photosynthesis is a complicated sequence consisting of many chemical reactions, but the net effect can be written quite simply as

$$6\ CO_2 + 6\ H_2O \rightarrow C_6H_{12}O_6 + 6\ O_2$$

where visible sunlight provides the added energy needed for the processes to proceed. Carbon dioxide and liquid water are readily available on Earth, and there is abundant sunlight to provide a source of energy. The organic molecule $C_6H_{12}O_6$ becomes part of the substance of the plant. Finally, gaseous O_2 is released as a byproduct, and this eventually accumulated to become almost 21% of the atmosphere by volume.

If photosynthesis acting alone were the whole story, then the atmospheric O_2 amount would keep rising indefinitely. However, geologic evidence shows that the abundance of atmospheric O_2 has remained remarkably constant for the past several hundred million years. Therefore, something must be removing O_2 from the atmosphere at essentially the same rate as it is being added. In the modern world, the production of O_2 by photosynthesis is balanced, to a good approximation, by loss processes that also involve the presence of life. The mechanism by which animals, including humans, obtain energy is to consume organic molecules and break them down chemically using oxygen taken from the atmosphere. The process involved is respiration, and its net chemical effect is the reverse of photosynthesis. This is

$$C_6H_{12}O_6 + 6\ O_2 \rightarrow 6\ CO_2 + 6\ H_2O + \text{Energy}$$

If a student eats a sandwich, represented by $C_6H_{12}O_6$, and breathes O_2, a result is the release of energy that the student can use to go to class. This energy could be traced back through the cycle of respiration and photosynthesis to light from the sun. In a sense, all living things are solar powered since the ultimate origin of organic molecules is photosynthesis. The above net reaction also describes the process of decay in which organic molecules react with oxygen to produce carbon dioxide. This would include, for example, the decay of dead leaves and animals in a forest. The important point is that molecular oxygen is abundant in the Earth's atmosphere because of plants performing photosynthesis. The lifetime of an oxygen molecule in the modern atmosphere is on the order of 10^4 years before it is consumed in respiration or decay. Over much longer geologic times, chemical reactions with rocks and the burial of dead organic matter influence the atmospheric O_2 abundance, although these processes are much slower than those described above (Goody and Walker 1972). Over timescales of primary interest in this book, days to one century, it is acceptable to assume a balance between production of O_2 in photosynthesis and loss in respiration and decay.

Earth's nearest neighbors, Mars and Venus, have atmospheres composed almost entirely of carbon dioxide. In addition, CO_2 is abundant in volcanic emissions on Earth, yet it is still a trace gas despite the added modern-day source associated with burning fossil fuels. The small abundance of atmospheric CO_2 is coupled to the presence of liquid water. Most of the Earth is covered with oceans, and CO_2 dissolves in these waters. Biological processing in the ocean converts it into other forms of carbon, which eventually become incorporated into rocks. Without the large quantity of liquid water on Earth, there would be much more gaseous CO_2 in the atmosphere. Mars and Venus do not have oceans, and consequently gaseous CO_2 tends to remain in their atmospheres.

1.3   The State of the Atmosphere

To describe the state of the atmosphere at any location and time, what quantities does one need to specify? Three of the most fundamental variables are temperature, density, and pressure. Temperature measures the average speed of air molecules as they move about at random, colliding with their neighbors. Density refers either to the total number of molecules in a unit volume of air or to the total mass of air per unit volume. Finally, pressure measures the force exerted on a unit area exposed to the atmosphere. These three quantities are related to each other by an "equation of state" called the "ideal gas law," discussed in Section 1.8. Given any two of these variables, the third is easy to compute. Another significant variable is wind. To describe the motion of air at a point in space, one must specify both a wind speed and a direction. Next, air at any location contains a water vapor amount that varies over time. These five quantities, temperature, density, pressure, wind, and water vapor content, are the traditional variables of meteorology, and with the exception of density, they all appear on daily weather reports. However, to understand the underlying physical and chemical processes that determine the state of the atmosphere, some other items require consideration.

Solar radiation is the ultimate energy source for the Earth's climate system. Without absorption of solar energy, the equilibrium temperature of the planet would approach absolute zero. Solar energy also drives photosynthesis, which creates both organic material and molecular oxygen. Finally, solar radiation, especially in the ultraviolet part of the spectrum, is the driving energy for atmospheric chemistry. Sunlight may be the most fundamental of the quantities that determine the state of the Earth's atmosphere. However, "terrestrial radiation" has a similar level of importance. As described in Chapter 2, the solid Earth, oceans, and atmosphere all emit energy in the far infrared part of the electromagnetic spectrum. The human eye does not respond to this energy, but terrestrial radiation is present throughout the environment. Terrestrial radiation escaping from the top of the atmosphere acts to cool the planet since it constitutes a loss of energy from the system. Terrestrial radiation is also at the center of the greenhouse effect, which leads to a surface temperature on Earth warmer than would exist otherwise.

The final class of quantities needed to describe the state of the atmosphere includes the abundances of numerous trace gases as well as particles. Section 1.1 identified the two most publicized trace gases, ozone and carbon dioxide. One of the major functions of trace gases and particles involves altering the solar and terrestrial radiation fields. Consequently, they are major players in the theory of climate as well as being a focus of atmospheric chemistry.

Before attempting to understand temperature, density, and pressure, it is useful to visualize the atmosphere, or any gaseous medium, on the molecular scale. A gas is a collection of molecules that move about at random. Molecules in the gas phase move about individually; they do not stick together in clumps. It is convenient to imagine molecules as behaving like tiny golf balls, where the radius of a nitrogen or oxygen molecule is on the order of 10^{-10} m (Herzberg 1950; Present 1958). Table 1.3 summarizes some relevant characteristics of air molecules. Readers who need to review the metric system, the units used to measure various physical quantities, and their common abbreviations may refer to Section 1.10, Review Topic: Units and Systems of Measurement.

As viewed by a microscopic observer the size of a molecule, a volume of air is a dynamic medium. The tiny golf-ball molecules fly about at random, bumping into each other. After a collision, the molecules move off in differ-

TABLE 1.3 Characteristics of Air at the Molecular Scale

- Approximate radius of N_2 or O_2: 10^{-10} m
- Collision frequency (at the ground): $10^9 - 10^{10}$ s^{-1}
- Mean free path (at the ground): $10^{-8} - 10^{-7}$ m
- Number density (at the ground): 2.55×10^{25} m^{-3}

ent directions only to collide again. To large creatures such as humans, a volume of air appears to be sitting still, but at the small scale of individual molecules, this volume is akin to a demolition derby taking place in three dimensions, with molecules continually slamming into each other from all sides.

Given the preceding picture, the "collision frequency" in a gas is the number of collisions per second experienced by an individual molecule. Near the ground, each molecule collides with its neighbors several billion times per second. Furthermore, a molecule does not move very far before it collides yet again. The "mean free path" is defined as the average distance that a molecule travels between successive collisions. Near the ground, the mean free path is several times 10^{-8} m. The final item in Table 1.3 is the number of molecules per unit volume; at the ground, this is 2.55×10^{25} molecules m^{-3}. The molecules in a gas move around individually, but as a consequence of frequent collisions, a small mean free path, and a large density, a macroscopic volume of air exists as a single, identifiable entity. This allows one to think about a "parcel of air" that contains a fixed number of molecules and moves about as a cohesive unit.

1.4 Temperature: Definition and Altitude Dependence

With the information in Table 1.3, it is possible to define the concept of temperature. Consider a macroscopic volume of air. The molecules contained herein are moving about at random, although the volume as a whole is stationary. The speed of any one molecule changes each time it undergoes a collision, but for every molecule that slows down, another molecule speeds up, so that the average speed computed over all of molecules in the volume remains constant.

The concept of "kinetic energy" appears in a general physics course. An object with mass m and speed v has a kinetic energy given by $\frac{1}{2}mv^2$. What is so significant about this particular quantity? The laws of classical mechanics show that the combination $\frac{1}{2}mv^2$ is a conserved quantity when summed over a system of colliding "golf balls" on which no outside forces act. When two molecules collide, the total kinetic energy of the entire system before the collision is the same as the kinetic energy after the collision, even though one molecule speeds up and the other slows down. In a gas comprising a very large number of molecules per unit volume, the kinetic energy of any one molecule changes billions of times per second, but the kinetic energy of the entire system of molecules remains constant in time so long as energy is not added to or taken from the system. If a force field is present, such as an attraction between molecules, the sum of kinetic energy and another quantity, called "potential energy," is constant, but this complication need not be considered here.

Based on these considerations, the temperature of a gas is directly proportional to the average kinetic energy per molecule in the gas:

$$3/2\, kT = <\tfrac{1}{2}\, mv^2> = \text{Average Kinetic Energy per Molecule} \quad [1.4.1]$$

where the brackets indicate an average computed over a large number of molecules. In Eq. 1.4.1, T is the absolute temperature measured on the Kelvin scale, m is the mass of one molecule, and k is Boltzmann's constant, whose value is $k = 1.38 \times 10^{-23}$ J K^{-1}. Boltzmann's constant converts temperature into kinetic energy, where the value of k is fixed by requiring that the temperature difference between the melting and boiling points of liquid water be 100 K. As defined in Eq. 1.4.1, temperature measures kinetic energy contained in random molecular-scale motions. The instantaneous motion of a molecule can be decomposed into components along three perpendicular directions in space, and the theory of gases shows that the average kinetic energy per molecule directed along each axis is $\tfrac{1}{2}\, kT$. The factor of 3 in Eq. 1.4.1 arises from summing the energies over all three perpendicular axes. Instead of quoting a temperature, it would be equally informative to state the average kinetic energy of a molecule in the air. For example, instead of quoting a temperature of 293 K (about 68°F), one could say that the average kinetic energy of an air molecule is 6.07×10^{-21} J. It is unlikely that this terminology would be well received on the daily weather report, so the concept of temperature is here to stay.

For many applications, it is sufficient to assume that molecules act like tiny golf balls, although this is a simplification. A molecule of nitrogen or oxygen more closely resembles two golf balls connected by a spring, where each golf ball represents one of the bound atoms. The atoms can vibrate by alternately stretching and compressing the spring and can rotate as well. Temperature measures all of these forms of energy, vibrational and rotational, as well as translational kinetic energy. Furthermore, an important result called "equipartition of energy" shows that the same value of temperature applies to all of the forms of molecular-scale energy identified above.

Few people, aside from scientists, think in terms of absolute temperature, even though it is a very logical system. The review material in Section 1.10 summarizes three different scales in use for measuring temperature. These are the absolute or kelvin scale, the Celsius or centigrade scale, and the Fahrenheit scale.

The next topic is the temperature structure of the Earth's atmosphere, starting at the ground. In a globally averaged sense, the temperature of the Earth's surface is approximately $T = 288.2$ K (*U.S. Standard Atmosphere, 1976*). Chapter 2 addresses the physical mechanisms that lead to this result. These involve the balance between heating of the Earth by sunlight and cooling by loss of terrestrial energy to outer space, with an additional surface heating from the greenhouse effect. Of course, the temperature at any specific location, in general, differs from the global average. The most obvious temperature variation at the ground exists between the tropics and the North or South Poles. This latitudinal gradient arises from systematic variations in the amount of sunlight absorbed by a unit area on the Earth's surface as position varies from the equator to either pole. This behavior, described further in Section 2.9, results from the simple fact that the Earth is nearly spherical.

If the amount of solar energy absorbed per unit horizontal area uniquely determined the temperature at any location, weather forecasting would be a lot easier than it actually is. The motion of air around the planet introduces a major complication. Winds tend to move relatively warm air from low latitudes to high latitudes, while colder air makes the return trip toward the tropics. Ocean currents have a similar effect. Because of heat transport by motions of the air and oceans, temperature differences between the equator and both poles are much smaller than would exist under the influence of radiation alone.

Atmospheric temperature displays a systematic dependence on altitude. As height increases, through the first several kilometers above the ground, air temperature decreases, at least when averaged over a large horizontal area and periods of several days or longer. A readily seen manifestation of this is the persistence of snow on the tops of tall mountains all year long, even during summer. The solid line in Figure 1.1 depicts the average temperature profile at middle latitudes. Air temperature declines linearly with increasing altitude over the vertical region shown, from the ground up to an altitude of roughly 11 km. Note that altitude, the independent variable, is on the vertical axis in Figure 1.1. This is a standard format when plotting any atmospheric quantity as a function of height. The annually averaged temperature decreases from 288.2 K (59°F) at the ground to 216.7 K (−70°F) at an

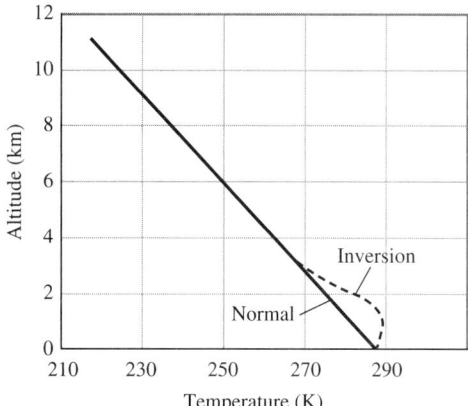

FIGURE 1.1   Vertical profiles of atmospheric temperature from the ground to an altitude of 11 km appropriate to middle latitudes. Solid line: A typical annually averaged profile. Dashed line: A smoothed profile in a temperature inversion adjacent to the ground.

altitude of 11 km. As a reference, 11 km is equivalent to 36,089 ft, the approximate altitude of a passenger aircraft at its highest point.

A quantity called the "lapse rate," Γ, describes the change in temperature with altitude. The definition used in this book is

$$\Gamma = -[\text{Change in Temperature}]/[\text{Change in Altitude}] = -\Delta T/\Delta z \quad [1.4.2]$$

where the Greek letter delta (Δ) means "change in." Note that a minus sign is placed on the right-hand side of Eq. 1.4.2. According to the sign convention adopted here, a decreasing temperature with increasing altitude corresponds to a positive lapse rate. The values in Figure 1.1 allow an evaluation of the average lapse rate between altitudes of 0 km and 11 km. This is

$$\Gamma = -\Delta T/\Delta z = -[T(11 \text{ km}) - T(0 \text{ km})]/[11 \text{ km} - 0 \text{ km}]$$

$$= -[216.7 \text{ K} - 288.2 \text{ K}]/[11 \text{ km}] = 6.5 \text{ K km}^{-1}$$

Chapter 3 considers the physical mechanisms that lead to this result. On average, the air temperature decreases by 6.5 K for every 1-km increase in height. Starting from a surface temperature of 288.2 K, this implies an average temperature equal to the freezing point of liquid water approximately 2.3 km above the ground. These numbers depict annual average conditions. The surface temperature at any location displays an annual cycle whose amplitude increases from low to high latitudes. For example, at latitude 45° north, typical surface temperatures range from 273 K in January to 296 K in July, with less pronounced differences occurring 11 km above the ground. This seasonal cycle implies a lapse rate in summer than exceeds that in winter (*U.S. Standard Atmosphere Supplements, 1966*).

On occasion, an event called a "temperature inversion" occurs. Here the temperature increases slightly with increasing altitude either in the region immediately adjacent to the ground or in a layer at higher altitudes. The dashed line in Figure 1.1 illustrates the former behavior. Chapter 3 shows that a temperature inversion inhibits small-scale vertical motions. Usually the lower atmosphere is constantly stirred by small-scale winds, but when these motions are suppressed, the air becomes stagnant. If the region where this occurs is a large city, automobile exhaust and industrial emissions remain in the area, and the result can be degraded air quality. The inversion effectively traps smoke and gaseous pollutants in a layer near the ground, and this appears as a brownish haze that sometimes hangs over the skyline of major cities on weekday mornings. This brown cloud comes from a buildup of pollutants associated with emissions from traffic, combined with a temperature inversion.

Temperature inversions can come about in two different ways. The first of these involves motion of two air masses. A shallow layer of relatively cold air might flow horizontally into a region near the ground, displacing warmer air there. Certain geographic regions are prone to the development of temperature inversions via the following sequence of events. A large body of water, like an ocean, stays cooler during summer than a land area adjacent to it. During the time span of one day, the ocean temperature remains essentially constant, while the nearby land heats up as the day proceeds. When air over the land mass warms, it tends to rise, while cooler air flows in from the ocean to replace it. Figure 1.2 illustrates these processes. Cool air is denser than warm air at the same pressure. The result of these air motions is a temperature inversion. Mountains on the right-hand side of Figure 1.2 act to restrict horizontal flow of the cooler air, keeping it over a specific region. If a city developed here, it would be prone to an air quality problem simply because of its geography.

The second common way to produce a temperature inversion is by "radiational cooling" during the nighttime hours. After sunset, heating of the ground by solar radiation is absent, and overnight the ground cools as it radiates terrestrial energy in the far infrared part of the spectrum. Terrestrial radiation is a subject for Chapter 2. The overall effect is that the ground cools faster than the air a short distance above it. As the ground cools, the air immediately adjacent to it also cools by being in contact with the surface. Over several hours of darkness, this process creates a temperature inversion that persists until after sunrise. After several hours of solar heating during the morning, the inversion dissipates. The classic text by Petterssen (1940) includes further details concerning the formation and characteristics of temperature inversions.

The temperature profile in Figure 1.1 extends only to 11 km. In the 19th century people were aware that atmospheric temperature declined with altitude, as was obvious from going up the side of a mountain. The presumption was that temperature continued to decrease upward until the atmosphere eventually faded into outer space, but this turned out to be

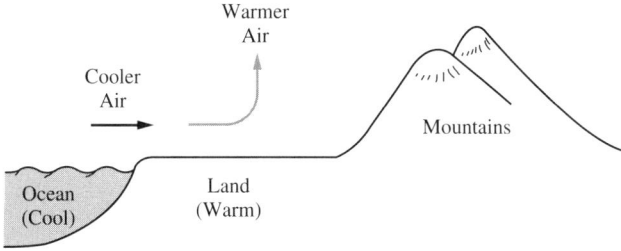

FIGURE 1.2   A geographic setting prone to experiencing temperature inversions: Air warmed by sunlight rises over land, to be replaced by cooler air flowing in from the ocean.

wrong. Humphreys (1964) gives an account of the early history of high-altitude atmospheric soundings. In 1898, Teisserenc de Bort began a series of upper air balloon soundings in France. The balloons carried thermometers that recorded air temperatures, and data obtained at the highest altitudes revealed some unexpected behavior. At heights around 10 to 11 km, the temperature stopped decreasing. Instead, it became almost constant with altitude and even showed signs of increasing slightly near the peak of the ascent.

Today, routine measurements of the atmospheric temperature profile exist from the ground up to around 60 km. These measurements are made by satellite-based sensors, which monitor infrared terrestrial energy escaping to space. In addition, balloon-borne measurements can reach altitudes near 30 km, while instruments carried on rockets extend to 100 km up and higher. This array of measurements, performed over many years, shows that the atmosphere consists of layers in which temperature decreases with altitude, sitting between layers where temperature increases with altitude. Figure 1.3 illustrates a typical atmospheric temperature profile at middle latitudes extending from the ground to an altitude of 120 km *(U.S. Standard Atmosphere, 1976)*.

With reference to Figure 1.3, there is a maximum temperature at the ground, as shown previously in Figure 1.1. Then there is a drop with a lapse rate of 6.5 K km^{-1} to a minimum near or somewhat above 10 to 11 km. A typical minimum temperature is near $T = 217$ K, as noted previously, but it

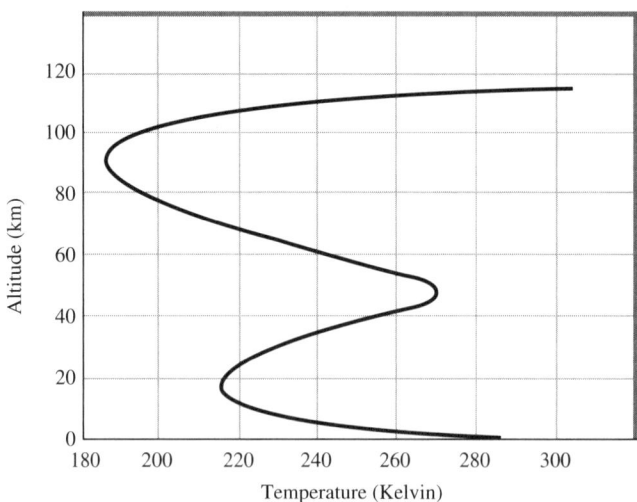

FIGURE 1.3   An atmospheric temperature profile appropriate to annually averaged conditions at middle latitudes for the altitude range from the ground to 120 km. [Source: *U.S. Standard Atmosphere, 1976*]

varies considerably with season and latitude. The situation changes going upward, where temperature increases with altitude to a maximum near 50 km, where a representative value is $T = 271$ K, almost to the melting point of ice. At still higher altitudes, temperature drops again. The coldest point in the atmosphere is at an altitude of 85 to 90 km, where the temperature is in the vicinity of 185 to 190 K. At still greater heights the air becomes warmer, and this trend continues to heights far above those shown in Figure 1.3. At an altitude of 200 km, the average temperature is typically 1000 to 1100 K, although at these altitudes the air is so thin that this region is effectively outer space.

1.5   Temperature: Nomenclature and Physical Mechanisms

It is convenient to adopt names for the various vertical layers of the atmosphere, and historically the temperature structure in Figure 1.3 has provided a basis for the nomenclature. The region closest to the ground, where air temperature decreases as altitude increases, is the "troposphere," and the specific altitude where the minimum temperature occurs is the "tropopause." At middle latitudes, the tropopause is near an altitude of 11 km, although numerous measurements show that the height of the tropopause varies with latitude. In the tropics, the tropopause can be 15 to 16 km high, while at latitudes 60° to 90° north or south, the temperature minimum might occur only 8 or 9 km above the ground.

Why should air temperature decrease with increasing altitude through the troposphere? The reasons for the decline are easy to identify, but a quantitative treatment of the atmosphere's thermal structure from first principles is a difficult undertaking. The qualitative explanation is as follows. Approximately half the incoming solar energy penetrates through the atmosphere to the ground and is absorbed there. This sunlight heats the Earth's surface, but the heating is uneven. Dark surfaces become warmer than lighter-colored ones, and air in contact with the ground picks up differing amounts of heat as a result. The discussion in Chapter 3 shows that a volume of warm air at the ground weighs less than the same volume of cooler air adjacent to it. As a consequence, the warmest parcels of air begin to rise by merit of their buoyancy.

As a parcel of air rises, its volume expands because there is less atmospheric weight pressing on it from above. Thermodynamics is a branch of classical physics that, among other things, describes the behavior of a gas when subjected to a variety of conditions. The First Law of Thermodynamics shows that when a volume of air expands, it also cools. Chapter 3 puts this result on a rigorous mathematical basis. Physically, a parcel of air expends energy when it expands against the surrounding atmosphere, and this energy comes at the expense of temperature, which measures the internal energy of molecules in the volume. In response to the change in internal energy, the rising, expanding air parcel cools. This process is called "expansional cooling," and it leads to colder air at higher altitudes. Conservation of mass is an important constraint on the system. A rising air parcel cannot leave a vacuum behind it at the ground; for every gram of air that rises, an equal mass of air has to descend. The described sequence now acts in reverse, and the sinking motion is accompanied by compression and warming. The net result of these rising and sinking motions is an air temperature that becomes colder with altitude. The technical name for this overturning of air in response to solar heating of the ground is "convection" or "thermal convection." Figure 1.4 illustrates these processes.

An observed atmospheric property, such as the lapse rate of tropospheric temperature, is often the result of several different physical mechanisms acting simultaneously. Upwelling air parcels carry water vapor with them, and much of this vapor condenses to liquid or freezes to ice to form clouds somewhere in the troposphere. When a gas condenses or freezes, energy is added to the surrounding air in a process called "latent heat release," a topic for Chapter 3. This new source of heating counteracts some of the cooling associated with rising motions, although cooling remains the dominant effect. As shown previously, the observed lapse rate for middle latitudes is $\Gamma = 6.5$ K km^{-1}. However, if thermal convection were the only process at work, the lapse rate would be 9.8 K km^{-1}. The difference between the two values, a heating of 3.3 K km^{-1}, arises from the latent heat released when wa-

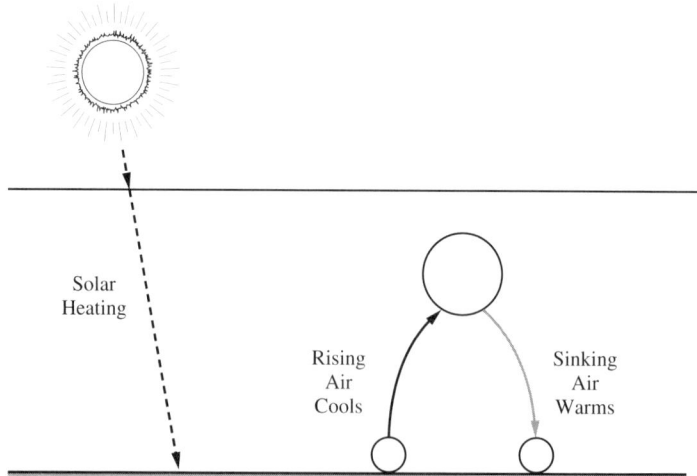

FIGURE 1.4   Processes that establish the atmospheric temperature profile in the troposphere: Rising air parcels expand and cool while sinking parcels are compressed and warm. The result is an atmospheric temperature that decreases as altitude increases.

ter vapor changes phase to liquid or ice. The observed temperature profile of the troposphere represents the combined effects of thermal convection and latent heat release.

The prefix *tropos* is Greek in origin, and it means changing or turning over in response to a stimulus. The troposphere is a region of changing weather, variable winds, and thermal convection. The stimulus that drives these phenomena is solar energy, and convective motions lead to a perpetual turning over of the atmosphere in response to this heating. Virtually all weather phenomena are confined to the troposphere. Common occurrences such as clouds, rain, and low-pressure systems, as well as extreme events like tornadoes and hurricanes, exist only in the lowest few kilometers of the atmosphere. The following list summarizes some characteristics of the troposphere.

Characteristics of the Troposphere

- Approximate altitude range: 0 to 11 km at middle latitudes, 0 to 15 or 16 km in the tropics, 0 to 8 or 9 km at high latitudes.
- Efficient convective motions, region of variable weather conditions.
- Contains approximately 80% of the atmosphere's total mass.
- Exchanges chemicals with the Earth's surface (land and oceans).

The preceding discussion addressed the altitude range and the presence of convective motions, but other traits of the troposphere are worth noting. First, as altitude increases the air gets progressively thinner. Consequently, most of the atmosphere's total mass, roughly 80%, lies in the troposphere. Second, proximity to the Earth's surface is a strong influence on the troposphere. Many chemicals enter the atmosphere via the ground. Water evaporates from oceans, and air pollutants are released in urban areas. In addition, processes in the troposphere are responsible for removing gases from the atmosphere. For example, numerous different trace gases dissolve in cloud water and fall out in rain.

Above the tropopause the air temperature, on average, starts to increase with altitude, and this continues up to about 50 km. Why should the temperature increase? The reason is that a source of heat exists at these altitudes, and it arises from the presence of the ozone layer. This region is the "stratosphere," and the specific altitude where the temperature reaches its maximum is called the "stratopause." The prefix *strato* means stable or stratified. The earlier discussion of temperature inversions near the ground noted that mixing of the atmosphere by vertical motions is suppressed when temperature increases with altitude. The air in a temperature inversion tends to be stable; it does not support efficient vertical mixing. By analogy, one can view the stratosphere as a global-scale temperature inversion sitting on top of the troposphere. The high degree of stability implies that material deposited in the stratosphere tends to remain there for an extended time, typically several years. At stratospheric altitudes, winds continue to spread material horizontally, but the stable temperature structure suppresses vertical mixing. In 1991, the volcanic eruption of Mount Pinatubo injected a large quantity of dust into the stratosphere, where some of it remained for several years. If the same amount of dust had been deposited in the troposphere, it would have returned to the ground under the action of efficient atmospheric mixing and precipitation within a few weeks. The inefficient vertical mixing in the stratosphere has implications for climate. During the early 1990s, volcanic dust reflected some of the incoming sunlight back into outer space, and this acted to cool the planet. In 1992, the Earth was slightly cooler than it would have been in the absence of the volcano's eruption.

The following list highlights some major characteristics of the stratosphere. The previous discussion addressed the altitude range and stability of the stratosphere. Next, the stratosphere is a region of active chemistry, and one of the results is the formation of the ozone layer. The ozone layer is important to life because it absorbs part of the sun's ultraviolet radiation, thereby preventing a biologically harmful component of sunlight from reaching the ground. In addition, when ozone absorbs ultraviolet radiation, energy from sunlight goes into heating the stratosphere, and this leads to the temperature maximum near 50 km in altitude. Today, there is a good overall understanding of the chemical processes that create and maintain the ozone layer, although there can be an occasional surprise. A prominent example is the "Antarctic ozone hole," a topic in Chapter 5. The prevailing theory of the mid 1980s did not predict this change in the south polar ozone layer. Its discovery via observations ultimately led to a revised understanding of chemical processes that destroy ozone in the stratosphere.

Characteristics of the Stratosphere

- Altitude range: tropopause to 50 km (approximately 11 to 50 km at middle latitudes).
- Weak vertical mixing and long residence times for gases and particles.
- Region of active chemistry: formation of the ozone layer.
- Absorption of solar ultraviolet radiation leading to temperature maximum near 50 km.

At altitudes above the stratopause, temperature decreases with altitude yet again. Most of the ozone in the Earth's atmosphere lies below 50 km, so there is little internal heating associated with this particular molecule at greater heights. The decline in temperature extends to 85 or 90 km. This region of decreasing temperature with height is the "mesosphere," and the specific altitude of the temperature minimum is the "mesopause." The prefix *meso* means middle or intermediate. Historically, scientists viewed the mesosphere as a middle ground between a lower atmosphere and an upper atmosphere. Today it is common to lump the stratosphere and mesosphere together and refer to them as the middle atmosphere. Atmospheric chemistry driven by intense ultraviolet sunlight is efficient in the mesosphere, although by far most of the Earth's ozone exists in the stratosphere, with only about 2% of the total being above 50 km.

Above 90 km, temperature increases with altitude, and the rise can be very steep. This high-altitude layer is the "thermosphere," where the prefix *thermo* obviously means that the region is hot. However, at these altitudes the air is tenuous. At an altitude of 95 km, which is very low in the thermosphere, the atmospheric density is about 1×10^{-6} times that at the ground. Artificial satellites can remain in Earth-orbit for extended periods at altitudes of 150 to 200 km, so for practical purposes the thermosphere is almost, but not quite, outer space. The molecules in the thermosphere are exposed to the shortest wavelengths of ultraviolet sunlight, and much of this energy is absorbed, leading to the very high temperatures. The energy absorbed in the thermosphere is miniscule compared to the total radiation provided by the sun, but when the atmosphere is extremely thin, a small amount of added energy can lead to a high temperature.

What happens when the atmosphere absorbs this solar energy, other than heating? Throughout much of the thermosphere, molecular nitrogen is still the most abundant molecule, so N_2 can serve as the example. The energy contained in short-wavelength ultraviolet sunlight can eject one of the electrons from the molecule. The notation for this process, called "ionization," is

$$N_2 + h\nu \rightarrow N_2^+ + e$$

where the symbol $h\nu$ refers to a photon of sunlight, defined in Chapter 2. The products of this ionization are a positively charged N_2^+ molecule, called a positive ion, and a negatively charged free electron, e. Energy from sunlight goes into breaking the electron free and into excess kinetic energy of both the N_2^+ and the electron. The kinetic energy is rapidly distributed among a large number of molecules via collisions and manifests itself as a high temperature. Because of ionization, positive ions and free electrons exist in the thermosphere, and for this reason, an alternate name for the thermosphere is the "ionosphere." After sunset, the abundance of electrically charged species at lower altitudes in the thermosphere decreases as electrons and positive ions combine to become neutral constituents. These diurnal changes in thermospheric ion densities influence radio communications. At night, an AM radio can pick up far-away signals that are undetectable during the day.

TABLE 1.4   Atmospheric Nomenclature Based on the Temperature Profile

Altitude Range (km)	Formal Name	Alternate Name
0–11 (middle latitudes)[a]	Troposphere	Lower atmosphere
11–50	Stratosphere	Middle atmosphere
50–90[b]	Mesosphere	Middle atmosphere
90–500	Thermosphere	Upper atmosphere
>500	Exosphere	Outer space

[a]The upper boundary of the troposphere varies with latitude, from 8 to 9 km in the polar regions to 15 to 16 km in the tropics.
[b]The upper boundary of the mesosphere is variable in the range 85 to 90 km.

The absence of a signal during the day arises from the dampening effect on radio waves of ions in the lower thermosphere.

The profile in Figure 1.3 stops at 120 km, but temperature continues to increase toward still higher altitudes. For instance, the temperature at 200 km varies from a low around 600 K to a high near 1500 K *(U.S. Standard Atmosphere, 1976),* and the time to cycle from the low to the high and back again is about 11 years. The long-term average temperature here is 1000 to 1100 K, but the amplitude of the 11-year cycle is quite large. What is taking place to cause this 11-year cycle in thermospheric temperature? The answer lies in the process of ionization discussed previously. The same mechanism that creates ions also adds energy to the thermosphere and leads to high temperatures. The very-short-wavelength solar energy that causes ionization varies with an 11-year period. The visible component of sunlight is nearly constant over periods of years to decades, but portions of the ultraviolet emission vary with an 11-year cycle. The heating rate of the thermosphere and therefore the temperature vary in the same way.

Strictly, there is no top to the atmosphere; the air becomes increasingly thin until it merges into outer space. Sometimes altitudes above 500 km go under the name "exosphere." The prefix *exo* means outside or external to; the exosphere is synonymous with outer space. This book does not consider the mesosphere, thermosphere, or exosphere further. The emphasis is on the troposphere and stratosphere, the two regions most directly linked to life at the Earth's surface. Table 1.4 summarizes the atmospheric nomenclature introduced in this section.

1.6 Gravity

The gravitational pull of the Earth is a major factor that determines the vertical structure of atmospheric density and pressure. Late in the 17th century, Isaac Newton formulated the Law of Universal Gravitation. Qualitatively this says that every object in the universe attracts every other object. For any two objects located anywhere in space, an attractive force acts to pull these objects together. This force is called "gravity," and the fundamental property of matter that determines the magnitude of the gravitational force is mass. Consider two objects of arbitrary shape separated by some distance r. These could be a star and a planet or two individual molecules; the only requirement is for the objects to have mass. Let the masses of the two objects be m_1 and m_2, respectively. Given this, Newton's Law of Universal Gravitation can be expressed quantitatively as

$$F = G\, m_1 m_2 / r^2 \qquad [1.6.1]$$

The attractive force (F) depends on the two masses, and it varies inversely as the square of the distance between them. Eq. 1.6.1 is not derived from anything more basic; its discovery came via creative insight, or inductive reasoning, motivated by observations. Newton proposed this mathematical relationship, and many experiments since have verified the inverse-square dependence. The quantity G is a constant, called the constant of universal gravitation, whose value is $G = 6.67 \times 10^{-11}$ N m^2 kg^{-2} (Hecht 1996). Quite appropriately, the metric unit of force is the newton (N), where 1 N is equivalent to 1 kg m s^{-2}.

How do objects respond to the gravitational force? Suppose the two objects are initially at rest. The attraction makes the two objects start to move toward each other. The motion is along the line that connects the two masses, and the objects accelerate toward each other, meaning that the velocities increase with time until the masses eventually collide. These concepts about the motion of masses in the presence of forces are contained in another of Newton's laws of motion, specifically Newton's Second Law of Motion. When a mass experiences a force, the mass accelerates in the direction of that force. Furthermore, acceleration is proportional to force, and mass is the proportionality constant that relates the two quantities. The quantitative statement of Newton's Second Law is force = mass × acceleration. In the case of gravity, the force F is given by Eq. 1.6.1. Both masses, m_1 and m_2, feel an attractive force of the same magnitude, and by Newton's Second Law, both of these masses accelerate according to

$$F = m_1 a_1 \text{ and } F = m_2 a_2 \qquad [1.6.2]$$

The force felt by each mass points toward the other, so the accelerations a_1 and a_2 are in opposite directions. Acceleration is the rate of change of velocity, accounting for both magnitude and direction. The magnitudes of the accelerations are

$$a_1 = dv_1 / dt \text{ and } a_2 = dv_2 / dt \qquad [1.6.3]$$

where v_1 and v_2 are the instantaneous speeds of the corresponding masses at time t, where the term *speed* refers to the magnitude of the velocity without regard for the direction.

The Earth is by far the most massive object that humans are close to, so the gravitational attraction of the Earth is usually the only important gravitational force in the immediate environment. Yet, according to Eq. 1.6.1, every object in the universe attracts every other object, although most of these forces are too small to merit attention. This is because either the masses are small or the distances are large. There are a few gravitational forces, other than the Earth's, that are sufficiently large to notice. The moon is not too distant, and the effect of its attractive force appears as tides, called lunar tides, in the ocean. Solar tides in the Earth's oceans also exist. The sun is a lot further away than the moon, so the $1/r^2$ term in Eq. 1.6.1 reduces the magnitude of the solar effect, but the mass of the sun is very large. The situation can get much more complicated. There are approximately nine planets in orbit around the sun, depending on the status assigned to Pluto, and by the Law of Universal Gravitation, an attractive force exists between all possible pairs of them. Over periods of thousands to hundreds of thousands of years, the gravitational pulls of other planets tend to tug on the Earth in a way that varies systematically over time and causes changes in the shape of the Earth's orbit around the sun. Milne et al. (1985) discuss this and other mechanisms that alter the amount of solar energy received by the planet.

Isaac Newton described the force of gravity in a simple equation, but why should gravity exist at all? Describing the action of gravity is one thing, but explaining why it exists is a lot harder. Why should an object sitting at one location attract another object located somewhere else? How does one object know that the other one is there? The origin of gravity involves abstract concepts, such as the structure of space and time and how mass distorts this structure (e.g., Davies 1977). These are not topics of study in atmospheric science, which simply accepts the fact that gravity exists and that Eq. 1.6.1 describes its effect.

The mathematical statement of the Law of Universal Gravitation specifies the gravitational force exerted on an object by the Earth. To analyze this further, let the mass m_1 in Eq. 1.6.1 refer to an object like a small rock close to the Earth or to the mass of a volume of air in the atmosphere. Then let the mass m_2 be the mass of the entire Earth, and to emphasize this, rename m_2 as M_E. The gravitational force between the object and the Earth is

$$F = G\, m_1 M_E / r^2 \qquad [1.6.4]$$

A conceptual difficulty in Eq. 1.6.4 involves the definition of r, the distance between the Earth and the object. Figure 1.5 depicts the situation. The Earth occupies a large volume, and all parts of the Earth are attracting the mass m_1. Furthermore, different portions of the Earth lie at different distances from

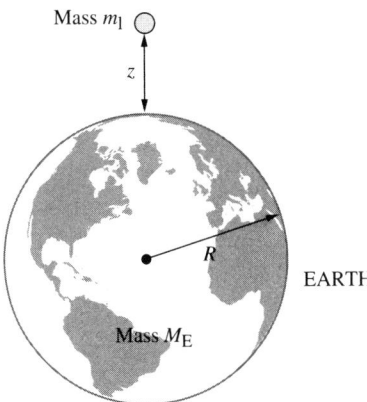

FIGURE 1.5   An object in the Earth's gravitational field: The object has mass m_1 and is a distance z above the Earth's surface. The mass and radius of the Earth are M_E and R, respectively.

m_1, so it is not immediately obvious what r should be. For example, the piece of the Earth immediately below an object sitting 10 km above the North Pole is a lot closer than a portion of the Earth near Australia or a location deep in the planet's interior, but all pieces of the Earth are attracting the mass m_1 simultaneously. With some mathematics, it is possible to show that the overall attractive force acts as if the entire mass of Earth is concentrated at the center of the planet. This can be derived rigorously provided the planet is a sphere and the distribution of mass inside the Earth depends only on distance from the center of the planet; that is, the Earth is spherically symmetric. Neither of these conditions is exactly true, but they are reasonable approximations.

The above result implies that r in the Law of Universal Gravitation should be the distance between the mass m_1 and the center of the Earth, or $r = R + z$, where R is the radius of the planet and z is the altitude of the mass above the Earth's surface. A perfect sphere with the same volume as the Earth would have $R = 6371$ km, so this is a reasonable value for the radius of the planet. Given this, the attractive force between the Earth and the mass is

$$F = G\, m_1 M_E/(R + z)^2 \qquad [1.6.5]$$

The focus here is on the troposphere and stratosphere, regions of the atmosphere where z is 50 km or less. In this case, $R + z$ in Eq. 1.6.5 is insignificantly larger than R alone, so it is acceptable to neglect z entirely and write

$$F = m_1(GM_E/R^2) \qquad [1.6.6]$$

The term in parentheses is a constant that depends only on the mass and radius of the Earth, and by comparison to Eq. 1.6.2 it has the dimensions of acceleration. It is convenient to define the "acceleration due to gravity" as $g = GM_E/R^2$. Newton's Law of Universal Gravitation applied to a mass m_1 subject to the Earth's gravitational force is now simply $F = m_1 g$.

If an object experiences the Earth's gravitational force, and no other influences are present, then the object accelerates toward the center of the planet, and the numerical value of this acceleration is g. This suggests an experiment to measure the value of g. All that needs to be done is to drop an object and measure its velocity as a function of time. Strictly, this experiment should be done in a vacuum where there is no added force from air resistance. Let the falling object have a downward velocity v_1 at time t. Newton's Second Law says

$$F = m_1 g = m_1 \, dv_1/dt \qquad [1.6.7]$$

Mass cancels from both sides of Eq. 1.6.7, so the acceleration of an object subject to the gravitational force is independent of the mass of that object. The resulting relationship, $dv_1/dt = g$, says that the velocity of a falling object increases linearly with time according to

$$v_1(t) = v_0 + g(t - t_0) \qquad [1.6.8]$$

where at an initial time t_0, the velocity is v_0. If the object is dropped at $t_0 = 0$, then $v_0 = 0$. Table 1.5 depicts the velocity of the falling mass as a function of time for the case where air resistance is negligible. Based on the values in Table 1.5, taken with Eq. 1.6.8, the acceleration due to gravity is $g = 9.807 \text{ m s}^{-2}$. The value of g deduced as described here includes a small contribution associated with the Earth's rotation. Values at the equator and poles are approximately 0.2% lower and higher, respectively, than those

TABLE 1.5 Velocity of an Object Accelerating in the Earth's Gravitational Field in the Absence of Air Resistance

Time t (s)	Velocity v (m s^{-1})
0	0
1	9.807
2	19.614 = 2 × 9.807
3	29.421 = 3 × 9.807
•	•
•	•
t	t × 9.807

stated (Hecht, 1996). Any ambiguity can be avoided by retaining only two significant figures and using $g = 9.8$ m s^{-2}.

The concept of weight, and specifically the weight of a volume of air, is important in an upcoming derivation. The weight of an object is the force exerted by the Earth's gravity on that object. This is

$$[\text{Weight of the Mass } m_1] = m_1 g \qquad [1.6.9]$$

An object has the same mass whether it is on Earth or in outer space. The mass depends on the number of protons, neutrons, and electrons that are present, and this is the same regardless of where in the universe the object happens to be. Weight, on the other hand, depends both on the object's mass and the gravitational field that the object experiences. The value of g differs from one planet and moon to another because it depends on the mass and radius of the body.

Eq. 1.6.7 describes the motion of an object in the Earth's gravitational field. Yet, Newton's Law of Universal Gravitation says that the small object, with mass m_1, attracts the Earth with the same force that the Earth attracts the object. Based on this reasoning, the small object should accelerate toward the Earth, and at the same time the Earth should accelerate toward the object. The motion of the Earth is described by

$$F = G\, m_1 M_E / R^2 = m_1 g = M_E\, dv_E / dt \qquad [1.6.10]$$

where v_E is the velocity of the entire planet as it accelerates toward the mass m_1. From Eq. 1.6.10, the acceleration of the Earth toward the object is $dv_E/dt = (m_1/M_E)g$. As an example, suppose that $m_1 = 1$ kg. The mass of the Earth, to one significant figure, is $M_E = 6 \times 10^{24}$ kg, so the acceleration of the Earth toward the object is $dv_E/dt = (1 \text{ kg}/6 \times 10^{24} \text{ kg})(9.8 \text{ m s}^{-2}) = 1.6 \times 10^{-24}$ m s^{-2}. With this acceleration, the Earth would move about 1 m in 350,000 years! Of course, the less massive object will crash into the planet after a brief period of acceleration. For any practical application, it is reasonable to assume that the Earth's motion is unaffected by the presence of small masses near it, but strictly speaking, this is an approximation.

There are two reasons why gravity is important in atmospheric science. The obvious one is that gravity holds the atmosphere close to the planet. But in addition, the force of gravity determines the vertical structure of density and pressure in the atmosphere, and this is the next topic.

1.7   Atmospheric Density and Pressure

There are two ways to measure the density of the atmosphere. The quantity used most frequently in this book is "number density." This is the total number of molecules per unit volume of air. The mathematical symbol is n, typically in units of m^{-3}. An alternate measure is "mass density," defined as the total mass per unit volume of air and usually denoted by ρ expressed in kg m^{-3}. As a representative number, the mass density of air at the ground is approximately 1.23 kg m^{-3}. The relationship between mass density and number density is

$$\rho = m\,n \qquad\qquad [1.7.1]$$

where m is the mean mass of an air molecule. For an atmosphere composed of 78% N$_2$, 21% O$_2$, and 1% Ar, the mean molecular mass is $m = 4.81 \times 10^{-26}$ kg, although the presence of water vapor would cause a slight reduction in this value.

A characteristic feature of planetary atmospheres is that density decreases as height increases. People who go to the top of a tall mountain can sense the decrease in atmospheric density by the lack of oxygen. There are fewer oxygen molecules per unit volume of air at high altitudes than at the ground, and it is necessary to breathe more frequently to get the same number of molecules flowing into the lungs. The reason for the decrease in density with altitude is that air is compressible. If one squeezes a volume of air, the volume gets smaller as the average distance between molecules shrinks. Compressibility is an important way in which a gas like the atmosphere differs from a liquid. The molecules in a liquid are already packed close together. Upon squeezing a liquid, it is not possible to cram the molecules much closer together than they already are. However, the molecules in a gas are relatively far apart. A gas is mostly empty space with molecules flying around at random and bumping into each other. It is easy to push the molecules of a gas closer together because of the large average distance between them.

Given the concept of compressibility, a simple analogy using the force due to gravity explains why air near the ground is denser than at high altitudes. Imagine four cardboard boxes, all of the same mass m, and these boxes are filled with a squishy material. In other words, the boxes are compressible. Now, suppose the boxes are stacked on top of each other. Let Box 1 be on the bottom; Box 2 is next, and so on to Box 4 on top. Figure 1.6 illustrates the situation. Each box feels the weight of all of the boxes above it as well as its own weight. Box 4 feels a gravitational force equal to its own weight. This is $F_4 = mg$. Box 3 experiences this same gravitational force because it has mass m, but Box 4 presses down from above. The total downward force on Box 3 is therefore $F_3 = 2\,mg$. This progression continues until Box 1 at the bottom of the stack experiences a force $F_1 = 4\,mg$. Since the boxes are compressible, the greater the force exerted, the more they are squashed. Near the bottom of the stack, the mass, m, gets crammed into a smaller volume than the same mass higher in the stack, or the mass per unit volume is greatest at the bottom. Therefore, the density of the boxes will decrease in a systematic

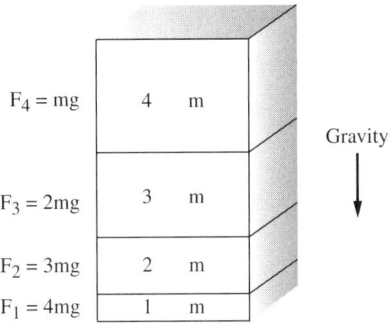

FIGURE 1.6   A stack of compressible boxes acted upon by the force of gravity: A box at any location in the stack is compressed by the weight of all of the boxes above it.

way as one moves up in the stack. The same situation exists in the atmosphere. The decrease in atmospheric density with height results from the compressibility of air combined with the downward force of gravity.

Pressure is closely related to density. Atmospheric pressure, denoted P, is defined as the force per unit area exerted on a surface exposed to the atmosphere. Accepted units are the pascal (Pa), where $1 \, Pa = 1 \, N \, m^{-2}$. In practice, other units appear frequently. These are the "bar," where $1 \, bar = 1 \times 10^5 \, Pa$, and especially the "millibar" (mb), where $1 \, mb = 1 \times 10^2 \, Pa$. This text prefers the millibar, where total atmospheric pressure at sea level is about 1013 mb.

To understand the origin of pressure, it is useful to focus on the molecules that make up a gas, in this case the atmosphere. As discussed previously, molecules are akin to tiny golf balls flying about in space. If a surface, like a wall, is in contact with the atmosphere, then air molecules will strike it and bounce off as depicted in Figure 1.7. Each individual collision exerts a force on the wall, and this is what pressure consists of. Pressure is the total force exerted on a unit area of the wall by all the molecules that strike it. Figure 1.7 shows only two molecules, but in the Earth's atmosphere each square meter of area experiences many billions of molecular collisions each second. As an analogy, suppose a professor has a bag full of golf balls and he throws one of them at a student sleeping in the class. The student feels a distinct force when the golf ball hits. Now if the professor could throw billions of golf balls per second, the student would tend to feel a continuous force. When the collisions become very frequent, the student experiences a force that seems constant in time instead of feeling each individual hit. The atmospheric pressure on a surface appears to be a continuous force because there are so many molecular collisions per second.

It is obvious that the atmospheric pressure and density are closely connected. The more molecules there are in a unit volume of air, the more collisions there will be with a surface placed in contact with this air.

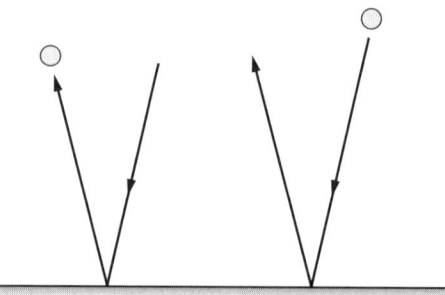

FIGURE 1.7   The origin of pressure when viewed at the molecular scale: Molecules exert a force on a surface upon collision. Pressure is the force per unit area exerted by all of the molecules that strike the surface.

Quantitatively, P is proportional to n, and this proportionality includes the concept of compressibility. If a layer of the atmosphere experiences more pressure from above, the number density gets larger. Increased pressure acts to pack the molecules closer together.

The definition of pressure in a gas is very general. It applies to the Earth's atmosphere as well as, for example, to a collection of molecules located in outer space. Furthermore, pressure can be exerted in any direction. A reference area placed in contact with a gas experiences the pressure associated with molecular collisions irrespective of the orientation of this area. However, if the gas resides in the gravitational field of a planet, there is another way to view pressure. Imagine a volume of air in a column that extends from the ground up to some very high altitude, and let the cross-sectional area of this column be A, as shown in Figure 1.8. The planet's gravity exerts a force on each molecule in this volume. The column experiences a downward gravitational force given by $F = Mg$, where M is the total mass of air in the column. Since pressure is defined as force per unit area, the atmospheric pressure on the ground is

$$P = Mg/A \qquad [1.7.2]$$

The weight of an object is the force exerted by gravity on that object, and in this example, the object is a column of air whose weight is Mg. Total atmospheric pressure at the ground is equal to the weight of a column of air with unit cross-sectional area extending from the ground to an altitude of, strictly, infinity. The same reasoning applies to a layer of finite depth. For example, the pressure exerted by a thin layer of air is equal to the weight of the layer per unit cross-sectional area. This way of viewing pressure will be useful in Section 1.9, which derives an equation to predict the altitude dependence of both pressure and number density throughout the atmosphere.

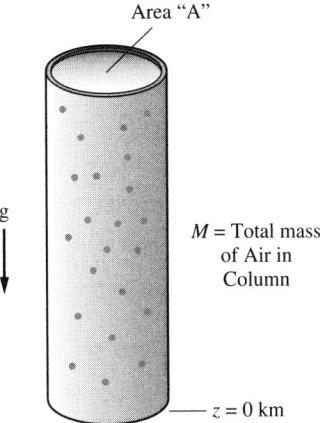

Area "A"

g

M = Total mass of Air in Column

$z = 0$ km

FIGURE 1.8 Atmospheric pressure in a gravitational field: Pressure is equal to the weight per unit cross-sectional area of the air above a specified level in the atmosphere.

The above interpretation of pressure can be criticized on the grounds that a volume of air is not a solid object but rather a collection of individual molecules separated by empty space (Bohren and Albrecht 1998). The pressure exerted on a surface by a gas clearly arises only from those molecules that actually collide with that surface. If the mean free path of air molecules near the ground is a miniscule fraction of a meter (see Table 1.3), how can atmospheric pressure at the Earth's surface include contributions from molecules that reside at altitudes several kilometers up? The missing element in this reasoning involves the role of collisions in a gas. Frequent collisions serve to transmit the gravitational force exerted on each molecule downward through the air column. The number density of the gas at the ground and, hence, the number of collisions per second with the surface are functions of this total downward force. This link leads to the equivalence between the gravitational interpretation of pressure in Eq. 1.7.2 and the molecular scale view in Figure 1.7.

1.8 The Ideal Gas Law

The connection between pressure and density is a logical consequence of compressibility. It is also true that the pressure and the temperature of a gas are closely related. Recall that temperature measures the average kinetic energy of molecules in the gas phase, so a high temperature means that the molecules have a large average speed. There are two ways to increase the pressure on a surface. First, a larger number of molecules can strike the surface per unit of time, corresponding to an increase in the number density of the gas. In this case, pressure increases because the surface is hit more frequently. The other way to increase pressure is to strike the surface with faster-moving molecules. Faster-moving molecules hit a surface with greater force. In this case, the surface is hit harder, and this corresponds to increasing temperature. To use the earlier analogy involving golf balls, the professor can increase the pressure on the sleeping student by throwing more golf balls per unit of time, analogous to increasing number density, or by throwing faster-moving golf balls, akin to increasing temperature. This reasoning illustrates that pressure is proportional to both number density and to temperature. These proportionalities are two pieces of the "ideal gas law." When combined, they imply the equality: $P = $ (constant) nT. A rigorous derivation shows that the constant in this expression is Boltzmann's constant, k, used in the definition of absolute temperature, Eq. 1.4.1. The final form of the ideal gas law is

$$P = nkT \qquad [1.8.1]$$

As stated previously, Boltzmann's constant is, in essence, a units conversion factor that transforms temperature into energy.

The ideal gas law was derived empirically several centuries ago based on laboratory experiments. A gas was heated, the volume was held constant, and changes in pressure were measured. Then the temperature was held constant, the volume was changed, and changes in pressure were measured. In these experiments, changing the volume of the gas was the way to change number density since the total number of molecules was fixed. The ideal gas law is a very old result, but thanks to Boltzmann's theoretical work in the late 19th century, it became possible to derive Eq. 1.8.1 based on a molecular model of a gas (Boltzmann 1964; Present 1958). To carry out the derivation, Boltzmann assumed that a gas consists of tiny golf balls that interact only upon collisions. At the time, this was a novel approach, but it yielded a result consistent with experiment.

1.9   Hydrostatic Balance

The section considers the physical mechanisms that determine the dependence of atmospheric pressure and number density on altitude and temperature. The compressibility of a gas explains the decrease in atmospheric density with altitude in a qualitative sense, but the concept of "hydrostatic balance" allows converting this into a quantitative model. Consider a thin layer of air at an altitude z, where the thickness of this layer is dz. Figure 1.9 summarizes the situation, where $dV = A\ dz$ is a volume increment with cross-sectional area A situated between z and $z + dz$. The number density

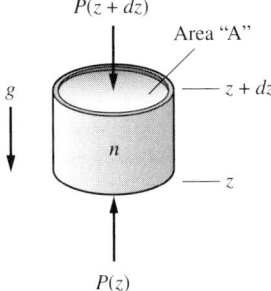

$P(z + dz)$

Area "A"

g

$z + dz$

n

z

$P(z)$

0

FIGURE 1.9   The forces on a layer of air with thickness dz at an altitude z: The balance of upward pressure, downward pressure, and gravity leads to the state of hydrostatic balance in which atmospheric pressure decreases as altitude increases.

of air in dV is n, and the layer is sufficiently thin that one can neglect changes in n over the infinitesimal distance dz. Finally, let m be the average molecular mass of air, where this average includes the proportions of N_2, O_2, Ar, and all other gases present.

The total mass of air in dV is $\mathfrak{M} = mn\, dV$, and the force of gravity acts to pull this mass $\mathfrak{M}$ toward the ground. The downward force F_1 is equal to the mass of the volume multiplied by the acceleration due to gravity:

$$F_1 = \mathfrak{M}\, g = mng\, A\, dz \qquad [1.9.1]$$

There is also a downward force associated with atmospheric pressure above the layer. Air molecules strike the top of the volume dV from above. More precisely, molecules at the upper boundary of this volume, altitude $z + dz$, are hit by molecules located just above the volume. This downward force is

$$F_2 = P(z + dz)\, A \qquad [1.9.2]$$

where $P(z + dz)$ is the downward atmospheric pressure at altitude $z + dz$. Based only on these two forces, the volume should accelerate toward the ground in accordance with Newton's Second Law. In practice, this does not happen because there is an upward force on the volume that balances the two downward forces. This force arises from the upward pressure exerted by molecules that strike the bottom of the layer from below. This force is

$$F_3 = P(z)\, A \qquad [1.9.3]$$

Newton's Second Law says that a mass subjected to a nonzero net force accelerates in the direction of that force. It is an observed fact that the atmosphere does not come crashing down to Earth, and neither does it fly away into outer space. For the volume of air to remain stationary, the three forces have to balance each other. Mathematically this is

$$F_3 - F_2 - F_1 = 0 \qquad [1.9.4]$$

where minus signs denote downward forces. Equations 1.9.1, 1.9.2, and 1.9.3 combined with Eq. 1.9.4 yield

$$P(z)\,A - P(z + dz)\,A - mng\,A\,dz = 0 \qquad [1.9.5]$$

or

$$P(z + dz) - P(z) = -\,mng\,dz \qquad [1.9.6]$$

The right-hand side of Eq. 1.9.6 is negative definite, so the pressure $P(z + dz)$ at the top of the layer must be smaller than the pressure $P(z)$ at the bottom.

Eq. 1.9.6 also results from the definition of downward pressure as the weight per unit area of a layer of air. The downward pressure at z is equal to the downward pressure at $z + dz$ plus the weight per unit area of air in the layer of thickness dz. This is $P(z) = P(z + dz) + W/A$, where W is the weight of the air in the volume dV, which is identical to the right-hand side of Eq. 1.9.1. The combination of the above relationships produces the same result as Eq. 1.9.6.

A Taylor Expansion allows rewriting the left-hand side of Eq. 1.9.6:

$$P(z + dz) = P(z) + [dP/dz]\,dz \qquad [1.9.7]$$

where the derivative is evaluated at altitude z. Use of Eq. 1.9.7 in 1.9.6 yields

$$dP/dz = -\,mng \qquad [1.9.8]$$

Equation 1.9.8 is the equation of hydrostatic balance.

The immediate goal is to derive an expression for the dependence of atmospheric pressure on altitude. The ideal gas law, Eq. 1.8.1, in the form $n = P/(kT)$ allows eliminating n from Eq. 1.9.8 in favor of P and T. This gives

$$dP/dz = -[mg/(kT)]\,P \qquad [1.9.9]$$

or

$$dP/P = -[mg/(kT)]\,dz \qquad [1.9.10]$$

Eqs. 1.9.9 and 1.9.10 still involve two unknowns, P and T, both of which depend on altitude. To proceed, one must assume that independent measurements of the temperature profile, T versus z, are available. Consider the dimensions in Eq. 1.9.10. The left-hand side is a pressure divided by a pressure, giving a unitless quantity. Clearly, the combination $kT/(mg)$ has the dimensions of length or height, and this quantity will prove useful.

An expression for atmospheric pressure as a function of altitude results from integrating both sides of Eq. 1.9.10 over z, but it is necessary to specify a boundary condition. At the ground, $z = 0$ km, and let $P = P_0$, where P_0 is the known surface pressure based on, for example, measurements. Then at any altitude z the corresponding pressure is $P(z)$. Similarly, let the measured temperatures at the ground and at any altitude z' be T_0 and $T(z')$ respectively. Integration of Eq. 1.9.10 upward over altitude from the ground to an arbitrary height z yields an expression whose left-hand side is a natural logarithm:

$$\int_0^z dP/P = \ln P(z) - \ln P_0 = \ln[P(z)/P_0] = \int_0^z dz' \, \{mg/[kT(z')]\} \quad [1.9.11]$$

The final solution for pressure is

$$P(z) = P_0 \exp\left[-\int_0^z dz' \, \{mg/[kT(z')]\}\right] \quad [1.9.12]$$

The ideal gas law applies at each individual point in a gas, so $P(z) = n(z)kT(z)$, and at the ground $P_0 = n_0 kT_0$, where n_0 and T_0 are the number density and temperature, respectively. The combination of Eq. 1.9.12 with the ideal gas law yields an expression for number density as a function of altitude:

$$n(z) = n_0 \, [T_0/T(z)] \int_0^z dz' \, \{mg/[kT(z')]\} \quad [1.9.13]$$

Strictly, the acceleration due to gravity varies with height, although over the depth of the troposphere and stratosphere, use of a constant value for g causes negligible error. For more accurate calculations, however, one should use the expression $g = g_0 \, [R/(R + z)^2]$ based on Eqs. 1.6.5 and 1.6.6. The value $g_0 = 9.8$ m s^{-2} applies at the ground, and R is the radius of the Earth.

Atmospheric temperature varies with altitude, so $T(z')$ should remain inside the integrals in Eqs. 1.9.12 and 1.9.13. However, if one assumes temperature to be independent of altitude so $T(z) = T_0 = T$ at all points, the results take on simpler forms. In this "isothermal atmosphere" the quantity

$$H = kT/(mg) \quad [1.9.14]$$

is independent of altitude, and the integrals over z' are trivially done to yield

$$P(z) = P_0 \exp(-z/H) \qquad [1.9.15]$$

$$n(z) = n_0 \exp(-z/H) \qquad [1.9.16]$$

Eqs. 1.9.15 and 1.9.16 describe the altitude dependence of pressure and number density in an isothermal atmosphere, where the quantity H is the "scale height." Over a vertical distance equal to H, atmospheric pressure and number density decrease by a factor of $e = 2.718$. Furthermore, as temperature increases, the scale height increases, and consequently pressure and density decline more slowly with altitude.

The results for an isothermal atmosphere are often sufficient for practical applications. One can split the atmosphere into layers each 1 or 2 km thick and assume that temperature is constant over the thickness of a single layer. This is equivalent to adopting an average temperature for each layer, where balloon soundings, for example, could provide the required values. Let T_i be the layer-averaged temperature of the ith layer, where layer $i = 1$ is adjacent to the ground, and let z_i be the top of layer i. The pressure at altitude z_1 is $P(z_1) = P_0 \exp(-z_1/H_1)$, where H_1 is the average scale height of layer 1, $H_1 = kT_1/(mg)$. The value $P(z_1)$ now becomes the pressure at the base of layer $i = 2$, so that $P(z_2) = P(z_1) \exp[-(z_2 - z_1)/H_2]$, where $H_2 = kT_2/(mg)$. This process of building up from a known pressure at the ground can continue to arbitrarily high altitudes to give, for any layer i,

$$P(z_i) = P(z_{i-1}) \exp[-(z_i - z_{i-1})/H_i] \qquad [1.9.17]$$

where the known scale height of layer i is $H_i = kT_i/(mg)$. An identical line of reasoning applies to the vertical profile of number density. Given the surface pressure and a temperature profile, it is possible to construct vertical profiles of atmospheric pressure and number density.

It is straightforward to compute the average scale height over the depth of the troposphere and stratosphere using measurements of atmospheric pressure or number density as a function of altitude. A plot of the logarithm of pressure or number density versus altitude produces nearly a straight line, and if the atmosphere were isothermal, the plot would be exactly this. Some representative values for total number density are, at the ground, $n_0 = 2.55 \times 10^{25}\ m^{-3}$ and $n(50\ km) = 2.14 \times 10^{22}\ m^{-3}$ at $z = 50$ km (*U.S. Standard Atmosphere, 1976*). The isothermal expression for number density, Eq. 1.9.16, can be solved for the scale height and evaluated numerically using the above values. This is

$$H = z/\ln[n_0/n(z)] = (50\ km)/\ln(2.55 \times 10^{25}\ m^{-3}/2.14 \times 10^{22}\ m^{-3}) = 7.06\ km$$

Hence, a typical average scale height over the altitude range of the troposphere and stratosphere is about 7 km.

Hydrostatic balance expresses the fact that the vertical structures of atmospheric pressure and density represent a balance between upward pressure, downward pressure, and the force of gravity in a compressible gas. The result is an exponential dependence on altitude. These concepts apply to the gaseous atmosphere of any planet. Different planets will have different scale heights because temperature, mean molecular mass, and the acceleration due to gravity change from one planet to another, but the principles are the same as used above. For the specific conditions found on Earth, the pressure and number density are, to a rough approximation, cut in half for every 5-km increase in altitude. For example, if the number density at the ground is one unit, then 5 km up it is approximately one half of a unit. It keeps dropping by a multiplicative factor of one half for every 5-km increase in altitude. Although this is an approximation, it produces values that lie within 20% of observations through the troposphere and stratosphere.

There is a subtle point buried in the above derivation, and it involves the mean molecular mass of the atmosphere, m in the definition of scale height. If, for example, the atmosphere is 78% N_2, 21% O_2, and 1% Ar, then $m = 0.78\ m_{N2} + 0.21\ m_{O2} + 0.01\ m_{Ar}$, where m_i refers to the mass of one molecule, or in the case of argon, one atom of constituent i. It is an observed fact that m is constant in altitude from the ground up to about 90 to 100 km, but it is not immediately obvious that this has to be true. As an example of a case where the mean molecular mass varies in the vertical, consider two different liquids, a very dense one like mercury and a relatively light one like water. Suppose these two liquids are poured together into a container. The denser liquid, mercury, sinks to the bottom and the lighter one, water, floats on top, so the mean molecular mass of the liquid decreases as altitude increases. This result seems reasonable on an intuitive basis, but the same type of reasoning applied to the atmosphere suggests that the heavier gases should collect near the ground. These would be Ar and O_2. Then N_2, being the lightest gas, would be relatively more abundant as altitude increases. The result is that m would decrease with altitude, as the lighter gas effectively floats on top of the heavier gases. While this may appear logical, atmospheric observations contradict it. Indeed, they show that m is constant throughout the lowest 90 to 100 km of the atmosphere. The reason for this is that small-scale motions are constantly mixing the atmosphere. This perpetual stirring keeps the proportions of the various major gases constant and results in a mean molecular mass that is independent of altitude. This mixing overpowers the tendency for the heavier gases, Ar and O_2, to sink toward the ground and for the lighter gas, N_2, to float on top.

The situation changes at altitudes above 100 km. When atmospheric number density is small, mixing by bulk motions becomes inefficient. This occurs in the region 90 to 100 km above the ground, at a level sometimes called the "turbopause," corresponding to the cessation of rapid mixing. At a sufficiently high altitude, each type of atmospheric constituent assumes its own individual scale height determined by its molecular or atomic mass. For example, molecular nitrogen assumes the scale height $H_{N2} = kT/(m_{N2}\, g)$. Similarly, the vertical profiles of Ar and O_2 will vary according to their individual scale heights. For this reason, the lighter gases become relatively more abundant as altitude increases above the turbopause. In the troposphere, atomic hydrogen is virtually nonexistent. However, at the highest levels in the thermosphere, it becomes a major gas because of its very small atomic mass.

1.10 Review Topic: Units and Systems of Measurement

To define the state of the Earth's atmosphere, or any other physical system, one must specify numerical values of various quantities. These are not simply numbers; they have "dimensions" or "units" as well. All physical quantities are expressed by stating a number and the associated units. The three fundamental dimensions are distance, time, and mass. There is no unique approach to selecting a set of units in which to measure these quantities, but once a system is adopted, it is essential to stick with it. If different people chose to measure distance, time, and mass using different units, it would be virtually impossible to exchange quantitative information.

The standard set of units used in the physical sciences is the System Internationale (SI), based on the metric system. Here the units of distance, mass, and time are the meter (m), kilogram (kg), and second (s), respectively ("mks" units). A variation of this uses the centimeter (cm), gram (g), and second ("cgs" units). Although the meter is the accepted measure of length, use of other units can be desirable, depending on the situation. For example, nanometers (nm) or micrometers (μm) are convenient when considering the wavelength of light, while the kilometer is reasonable for measuring height in the atmosphere. Relationships between various metric units are as follows: 1 nm = 1×10^{-9} m, 1 μm = 1×10^{-6} m, 1 millimeter (mm) = 1×10^{-3} m, 1 cm = 1×10^{-2} m, and 1 km = 1×10^{3} m. The prefixes nano-, micro-, milli-, centi-, and kilo- always imply multiplying the reference unit by 1×10^{-9}, 1×10^{-6}, 1×10^{-3}, 1×10^{-2}, and 1×10^{3}, respectively. For example, in the case of mass, 1 kg = 1×10^{3} g.

Conversions between the units of length and mass all involve various powers of 10. In the case of time, the relationships are less systematic. These are 1 minute = 6×10^{1} s, 1 hour (hr) = 3.6×10^{3} s, and 1 day = 8.64×10^{4} s. With the exception of temperature, all quantities in the physical sciences can be expressed as combinations of the three basic units of distance, time, and mass. Some common physical quantities and their units appear in Table 1.6.

Of the three scales in use for measuring temperature, the absolute or kelvin scale has the most straightforward link to the microscopic properties of matter. Absolute temperature is proportional to the average kinetic energy per molecule, as expressed in Eq. 1.4.1, where Boltzmann's constant provides the connection between absolute temperature and energy ex-

TABLE 1.6   A Review of Dimensions and Metric Units for Common Physical Quantities

Quantity (dimensions)	Typical Metric Units
Distance	meter (m)
	1 kilometer (km) = 1×10^3 m
	1 centimeter (cm) = 1×10^{-2} m
	1 millimeter (mm) = 1×10^{-3} m
	1 micrometer (μm) = 1×10^{-6} m
	1 nanometer (nm) = 1×10^{-9} m
Mass	gram (g)
	1 kilogram (kg) = 1×10^3 g
	1 milligram (mg) = 1×10^{-3} g
	1 microgram (mg) = 1×10^{-6} g
Time	second (s)
Area (distance^2)	m^2
	cm^2
Volume (distance^3)	m^3
	cm^3
Velocity (distance time^{-1})	m s^{-1}
	cm s^{-1}
Acceleration (velocity time^{-1})	m s^{-2}
	cm s^{-2}
Number density (volume^{-1})	m^{-3}
	cm^{-3}
Mass density (mass volume^{-1})	kg m^{-3}
	g cm^{-3}
Force (mass $\times$ acceleration)	1 newton (N) = 1 kg m s^{-2}
	1 dyne (d) = 1 g cm s^{-2}
Pressure (force area^{-1})	1 pascal (Pa) = 1 N m^{-2}
	1 bar (b) = 1×10^5 Pa
	1 millibar (mb) = 1×10^2 Pa
Energy (force $\times$ distance)	1 joule (J) = 1 N m
	1 erg = 1 d cm
Power (energy time^{-1})	1 watt (W) = 1 J s^{-1}
Energy Flux (energy area^{-1} time^{-1})	W m^{-2}

pressed in metric units. The commonly used Celsius or centigrade scale is based on the properties of liquid water, where 0°C corresponds to the freezing point and 100°C corresponds to the boiling point at a reference atmospheric pressure. The analogous temperatures on the Fahrenheit scale are 32°F and 212°F, respectively. One unit on the kelvin scale is equivalent in energy to one degree on the Celsius scale. The only difference is the shift in zero points. The conversion from temperature on the kelvin scale, $T(K)$, to the Celsius scale, $T(C)$, is $T(K) = T(C) + 273.15$. A one-degree shift on the Celsius or kelvin scales corresponds to a greater energy change than one degree on the Fahrenheit scale. A degree Fahrenheit represents a smaller change in energy than does a degree Celsius. The conversion is $T(C) = (5/9)[T(F) - 32]$, where $T(F)$ is temperature in degrees Fahrenheit.

1.11 Exercises

1. The definition of absolute temperature in Eq. 1.4.1 involves $<v^2>$, the average value of the square of the molecular speeds in a gas. This is not the same as $<v>^2$, the average value of the molecular speeds squared. To demonstrate this inequality, consider 10 cars moving along a highway. One car is moving at 20 miles per hour (mph). Eight cars are moving at 30 mph and one car is moving at 50 mph. Compute the average speed of the 10 cars, $<v>$, and square it to obtain $<v>^2$. Now compute the average value of the squares of the speeds, $<v^2>$. How does $<v>^2$ compare to $<v^2>$? The same principle applies to molecules moving at different speeds in a gas.

Exercises 2 through 5 below consider a gas with the following properties:

- The temperature is $T = 288$ K, and the number density is $n = 2.55 \times 10^{25}$ m^{-3}.
- All of the molecules are spheres each with a radius of $r = 2 \times 10^{-10}$ m.
- The gas is 100% composed of molecular oxygen whose mass is 32 grams mol^{-1}, where one mole of any gas contains 6.0225×10^{23} molecules. The mole is a convenient unit for measuring large numbers of molecules, and grams per mole is abbreviated "g mol^{-1}."
- Assume that $<v>^2 = 0.849<v^2>$, where v is the speed of a molecule and the brackets denote an average over a large number of molecules. This result is a consequence of the statistical distribution of molecular speeds in a gas where collisions are frequent (Present 1958).

2. Compute the average speed in m s^{-1} of a molecule in the gas described above. To get a feel for how fast the molecules are moving, an automobile moving at 65 mph has a speed of approximately 2.9×10^1 m s^{-1}. What is the ratio of the average speed of a molecule to that of the automobile?

3. The "collision frequency" (Q) is defined as the number of collisions per second experienced by one molecule in a gas. A rigorous derivation shows that

$$Q = 4\pi \, r^2 <v> n$$

with the units collisions per second (s^{-1}), where r, v, and n are as defined previously and $<v>$ is the average speed of a molecule from problem 2.

a. Use the numerical values given above to compute the collision frequency of molecules in the gas. As a reference point, in 2005, between 6 billion and 7 billion ($6–7 \times 10^9$) people were alive on the Earth. If each person bumped into all of the other people every second, the collision frequency of one human would be similar to that of a molecule in the gas.

b. The collision frequency computed in problem 3a applies to a temperature of $T = 288$ K. Assume now that the temperature of the gas decreases, but the number density remains constant as the gas cools. To what value must the temperature fall for the collision frequency to become one half the value computed in problem 2a?

4. The mean free path (L) of a molecule is defined as the average distance moved between consecutive collisions. The mean free path is related to the average speed ($<v>$) and collision frequency (Q) by

$$<v> = LQ$$

a. Use the results from problems 2 and 3 to compute the mean free path (in m) of a molecule in the gas.

b. What would the number density of the gas have to be to produce a mean free path of 1 km? The correct result corresponds to the number density of the Earth's atmosphere near an altitude of 250 km, a typical altitude for the space shuttle's orbit.

5. Use the number density of the gas and the radius of a molecule to argue, quantitatively, that the gas is mostly empty space.

6. At the ground the total number density of the Earth's atmosphere is $n = 2.55 \times 10^{25}$ m^{-3}, and the average temperature is 288 K. The radius of the planet is $R = 6371$ km. Use these values to compute the total mass of the Earth's atmosphere.

7. Assume that the composition of the Earth's atmosphere is as follows: 78% of the total number density consists of N_2, 21% consists of O_2, and 1% consists of Ar. The molecular mass of N_2 is 28 g mol^{-1}, O_2 is 32 g mol^{-1}, and Ar is 40 g mol^{-1}. Use the information given here with the result from problem 6 to compute the total mass of molecular oxygen (O_2) in the Earth's atmosphere.

8. The exponential falloff of atmospheric density (Eq. 1.9.16) implies that the total number density never reaches zero regardless of how high altitude becomes. Suppose a satellite is in orbit at an altitude of 250 km above the Earth. Under normal conditions, there is very little atmosphere here, so there is negligible frictional drag on the satellite due to

TABLE 1.7 Measured Atmospheric Pressure (P) as a Function of Altitude (z) on a Distant Planet

z(km)	P(mb)
8	268.0
6	362.0
4	479.4
2	624.0
0	800.0

collisions with atmospheric molecules, and the orbit of the satellite is stable. However, after a large explosion on the sun (called a "solar flare"), energetic solar particles heat the Earth's uppermost atmosphere, leading to a large increase in temperature. Under these disturbed conditions, an increase in the frictional drag on the satellite develops, and the orbit of the spacecraft begins to decay. Explain the reasons for this observed behavior using the concept of hydrostatic balance.

9. A spacecraft goes into orbit around a distant planet and releases a landing module (LM) that descends through the atmosphere toward the planet's surface. A laser beam from the LM repeatedly bounces off the surface to measure altitude. The LM carries sensors designed to measure the temperature and pressure of the planet's atmosphere as functions of altitude. Unfortunately, the temperature sensor fails to operate, so that the only readings available are pressure and altitude. Table 1.7 presents the data collected as the LM descends, where $z = 0$ km refers to the planet's surface. The atmosphere is composed of 100% CO_2, and the acceleration due to gravity is $g = 6.5$ m s^{-2}. The mass of CO_2 is 44 g mol^{-1}.

a. Estimate the lapse rate of temperature in the planet's atmosphere.

b. Estimate the surface temperature of the planet. Will water on the planet's surface be liquid or ice?

1.12 References

Bohren, C. F., and B. A. Albrecht. *Atmospheric Thermodynamics*. New York: Oxford University Press, 1998.

Boltzmann, L. *Lectures on Gas Theory*. Translated by S. G. Brush. Los Angeles: University of California Press, 1964.

Davies, P. C. W. *Space and Time in the Modern Universe*. Cambridge: Cambridge University Press, 1977.

Goody, R. M., and J. C. G. Walker. *Atmospheres*. Englewood Cliffs, N.J.: Prentice-Hall, 1972.

Hecht, E. *Physics: Calculus*. Pacific Grove, Calif.: Brooks Cole Publishing, 1996.

Herzberg, G. *Molecular Spectra and Molecular Structure I. Spectra of Diatomic Molecules*. New York: Van Nostrand Reinhold, 1950.

Holland, H. D. *The Chemical Evolution of the Atmosphere and Oceans*. Princeton, N.J.: Princeton University Press, 1984.

Humphreys, W. J. *Physics of the Air*. New York: Dover Publications, 1964.

Milne, D., D. Raup, J. Billingham, K. Niklaus, and K. Padian. *The Evolution of Complex and Higher Organisms*. Washington, D.C.: National Aeronautics and Space Administration, 1985.

Petterssen, S. *Weather Analysis and Forecasting*. New York: McGraw-Hill, 1940.

Present, R. D. *Kinetic Theory of Gases*. New York: McGraw-Hill, 1958.

Salstein, D. A. "Mean Properties of the Atmosphere." *Composition, Chemistry and Climate of the Atmosphere,* edited by H. B. Singh, 19–49. New York: Van Nostrand Reinhold, 1995.

U.S. Standard Atmosphere Supplements, 1966. Washington, D.C.: Government Printing Office, 1966.

U.S. Standard Atmosphere, 1976. Washington, D.C.: Government Printing Office, 1976.

Walker, J. C. G. *Evolution of the Atmosphere*. New York: Macmillan, 1977.

Chapter 2

Solar and Terrestrial Radiation:
The Atmospheric Energy Balance

INTRODUCTION

Light, or electromagnetic radiation, is a fundamental way in which energy propagates through space. Processes taking place in the interior of the sun and the distance between the sun and the Earth determine the magnitude of the radiant energy incident on the planet. A fraction of the incoming solar energy is absorbed and heats the Earth-plus-atmosphere system. A compensating cooling occurs in the form of infrared terrestrial radiation emitted by the planet and atmosphere, the magnitude of which is a sensitive function of temperature described by the Stefan–Boltzmann Radiation Law. The effective radiating temperature of the Earth-plus-atmosphere system is the value at which the terrestrial energy lost to space is balanced by the solar energy absorbed, leading to the state of radiative equilibrium.

The greenhouse effect provides an additional heating of the Earth's surface. Atmospheric gases such as carbon dioxide and water vapor absorb terrestrial radiation upwelling from the ground and then reemit it both upward and downward. The portion that reaches the ground provides the extra heating required to produce the observed surface temperature, which is about 33 K higher than the effective radiating temperature. Although radiative equilibrium applies in a globally integrated sense, the spherical geometry of the Earth and the inclination of the equator to the plane of the orbit around the sun lead to a systematic annual variation in solar energy received at any fixed latitude. The consequence is the observed seasonal cycle.

2.1 Conceptual Models for Light

Sunlight is the ultimate energy source for the Earth. Absorption of solar energy by the solid ground, oceans, and atmosphere drives the climate system, and the sun's heat creates temperatures adequate to support life. Some of the solar energy evaporates liquid water from the oceans, initiating the global water cycle that creates clouds and precipitation. Visible sunlight drives photosynthesis, which produces molecular oxygen, and ultraviolet solar radiation is the energy source for chemical reactions in the atmosphere.

The sun is not the Earth's only source of energy. Energy from other stars provides a very small, insignificant heating, while another source is geothermal heating. The interior of the Earth is hot, and the decay of radioisotopes produces heat. Some of this energy appears at the planet's surface in the form of volcanic heat and geysers. However, for purposes of this chapter, the sun is the only energy source that need be considered.

The sun emits energy in the form of "electromagnetic radiation," which encompasses visible light as well as other regions of the spectrum to which the human eye is not sensitive. Light is a fundamental way in which energy moves from one location to another. Rigorous definitions of light are based on a branch of physics called electromagnetic theory as well as on more recent concepts based on quantum theory. There are two conceptual models for light, neither of which is complete if taken alone. One model is the "wave description" and the other is the "particle description"; the latter is the newer of the two. This second model views a beam of light as consisting of a stream of tiny particle-like entities called "photons" moving through space. Photons possess some unusual attributes. First, in empty space they all travel at the same speed, called the speed of light, usually denoted by c, where to two significant figures $c = 3.0 \times 10^8$ m s^{-1} in a vacuum (Hecht 1996). At this speed, it takes a photon approximately 8.3 minutes to move from the sun to the Earth. The following list summarizes some of the distinguishing properties of photons.

Some Properties of Photons

- All move at the fixed speed $c = 3.0 \times 10^8$ m s^{-1} in a vacuum.
- A photon has a mass of zero.
- A photon is a "bundle of energy" flying through space.
- The energy of a photon is equivalent to the color of light.

Although it is convenient to visualize photons as akin to particles, reality is more complicated than this picture suggests. A photon is not a solid particle, so it is fundamentally different from an electron or a proton. A photon has no mass, and it is effectively a bundle of pure energy. This may be a vague description, but it is the best qualitative statement one can give. Photons are bundles of energy that fly through a vacuum at a universal speed. There is no such thing as a photon that sits still; if a photon exists at all, it moves at the speed of light. All photons may have a mass of zero and move at the same speed, but they are not identical. Photons exist over a range of energies, and the human eye and brain interpret photons with different energies as having different colors. Energy per photon and the color of light refer to the same concept. Of course, energy per photon is the true physical quantity, whereas color is a less rigorous concept that is related to how the eye and brain respond to light. In addition, the human eye reacts only to photons in a limited range of energy, appropriately called the visible region.

The particle description of light was developed early in the twentieth century, and among other things, it provides a conceptual framework for explaining the interaction of light with atoms and molecules. Imagine an experiment in which a beam of light passes through a gas. For concreteness, assume that the gas consists of oxygen atoms. One detector measures the spectrum of light incident on the gas, and a second detector measures the spectrum transmitted through the gas. The term "spectrum" refers to the fact that the light source emits photons spread over a range of energies. Let the incident spectrum be as depicted by the solid line in Figure 2.1. Photons spread over a range of energies are present initially, and the term "flux" refers to the number of photons in a narrow interval of wavelength that cross unit area in unit time. The light transmitted through the gas would appear as shown by the dashed lines in Figure 2.1, where bites are taken out of the spectrum at discrete energies. The atoms absorb only photons with certain well-defined energies. Observations like that described here led to the idea that atoms and molecules exist only in certain discrete states that correspond to specific energies. Herzberg (1945) has given a comprehensive review of these and other aspects of atomic structure.

Consider an atom consisting of a positively charged nucleus with negatively charged electrons moving in orbits around that nucleus. An energy state of the atom corresponds to a specific arrangement of these electrons. There is a set of "quantized" energy states available to the atom. For exam-

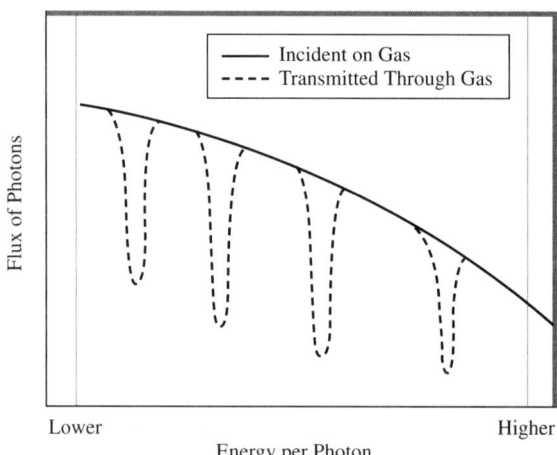

FIGURE 2.1   Absorption of a beam of light by a gas composed of atoms. The atoms absorb only those photons whose energies match the spacing between discrete atomic energy levels.

ple, with reference to Figure 2.2, an electron can occupy an orbit whose radius is r_1 or r_2, but no other orbit between r_1 or at r_2 can exist. Energy needs to be added to the atom to move an electron from r_1 to r_2, and a photon of light can provide this energy. An atom can jump from state 1 to state 2 by absorbing, and thereby destroying, a photon whose energy matches the energy difference between the states. A correspondence exists between the discrete energy states in an atom and the energies of photons that this atom will absorb. One atom can absorb one photon if there is a match between an energy level spacing in the atom and the energy of the photon. The concept that light comes in discrete bundles of energy fits nicely with the existence of these quantized energy states in atoms and molecules.

At this point, a concept from the wave theory of light is useful. The energy per photon is a fundamental property of a beam of light, but a concept called "wavelength" is commonly used in its place. The energy of a photon is inversely related to the associated wavelength, where the relationship is

$$E = hc / \lambda \qquad\qquad [2.1.1]$$

Here E is the energy of a photon, λ is the wavelength, c is the speed of light, and h is "Planck's constant," whose value is 6.626×10^{-27} erg s or 6.626×10^{-34} J s (Hecht, 1996). Eq. 2.1.1 implies the existence of an entity, today called a photon, whose energy is E. When first proposed by Max Planck near the turn of the twentieth century, this constituted a profound hypothesis concerning the nature of electromagnetic radiation, which previously was treated purely as a wave phenomenon (Planck 1959).

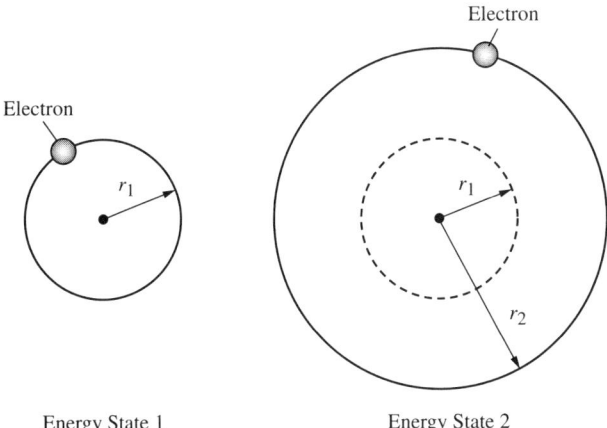

Energy State 1   Energy State 2

FIGURE 2.2 The electrons in an atom occupy only specific, discrete orbits. An electron can exist in an orbit with radius r_1 or r_2, but no additional orbits between r_1 and r_2 exist. Different energy states of an atom correspond to different arrangements of electrons in the allowed orbits.

The wave theory of light attaches a physical meaning to wavelength. As an analogy, consider water waves on the surface of a lake, depicted in Figure 2.3. One wavelength is the distance from one crest to the next. A boat moving in the middle of the lake disturbs the surface of the water and creates the wave that moves away from its source and propagates to the shore. A water wave consists of the water surface moving up and down, with this oscillation propagating across the surface of the lake. This picture is straightforward for water waves, but when a beam of light is described as a wave, exactly what is it that is waving? What physical quantity is it that oscillates and propagates through space? The answer is electric and magnetic fields, and hence the name electromagnetic radiation. Just as the motion of the boat created water waves, the motion of "electrical charges" creates electromagnetic waves. To pursue this further, one would have to define the concept of electrical charge as well as that of electric and magnetic fields. This would lead to rigorous mathematical descriptions that ultimately predict the existence of electromagnetic waves that propagate through space at the speed of light (e.g., Feynman et al. 1964). For the applications in this text, it is often most useful to adopt the particle description of light, and in this view wavelength is simply an alternate way to measure the energy of a photon.

It is useful to introduce some terminology to describe the electromagnetic spectrum and to facilitate the upcoming discussion of sunlight. Table 2.1 summarizes the major subdivisions of the electromagnetic spectrum. The nanometer, where 1 nm $= 1 \times 10^{-9}$ m, is a convenient unit for measuring wavelength, although in the infrared, micrometers (μm), where 1 μm $= 1 \times 10^{-6}$ m, are used frequently as well. Ultraviolet, visible, and infrared are the most significant spectral regions for understanding processes in the Earth's

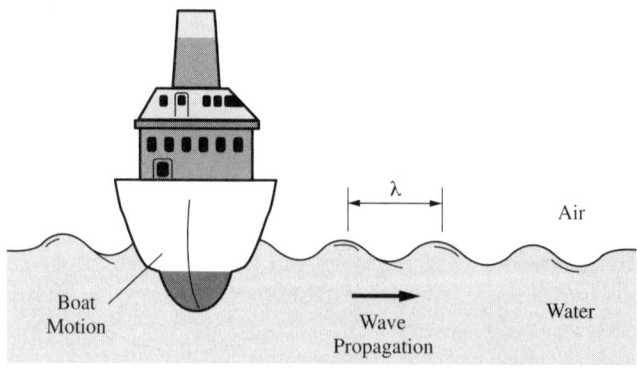

λ = One Wavelength

FIGURE 2.3   Water waves on the surface of a lake. A disturbance in the form of a moving boat creates the waves that propagate toward the shore. By analogy, the motion of electrical charges creates electromagnetic waves that propagate through space.

TABLE 2.1   Regions of the Electromagnetic Spectrum

Wavelength (nm)	Name	Energy Per Photon
Shorter than 0.1 nm	Gamma Rays	Very High
0.1–10 nm	X-Rays	High
10–400 nm	Ultraviolet	High/Medium
400–780 nm	Visible	Medium/Low
780–20,000 nm	Infrared	Low
Longer than 20,000 nm	Microwaves, Radio Waves	Very Low

troposphere and stratosphere. The ultraviolet region is important because it drives atmospheric chemistry and is associated with a variety of biological effects when it reaches the ground. The visible and infrared regions are significant because they heat the Earth's surface and the lowest parts of the troposphere, thereby exerting a major influence on the temperature of the planet. The human eye evolved so it responds to the visible part of the spectrum. The atmosphere is relatively transparent in the visible, so a substantial fraction of the light at 400 to 780 nm reaches the ground. Subdivisions of the visible correspond to colors, from red at the long wavelength boundary through orange, yellow, green, and blue to violet at the short end. Wavelength and energy per photon are much more rigorous ways to describe light than color, which is qualitative and applies only to the visible.

This terminology is the basis for describing the energy emitted by the sun. The sun emits photons spread over the broad range of wavelengths defined in Table 2.1. A plot of solar energy as a function of wavelength produces a "solar spectrum," as depicted in Figure 2.4. The vertical scale is the energy per unit area per unit time per unit wavelength interval received at the top of the Earth's atmosphere, where the reference area is oriented to face the sun. The technical name for this quantity is the "extraterrestrial solar spectral irradiance," and typical units are watts per square meter per nanometer of wavelength.

At the shortest wavelengths, gamma rays and X-rays are very minor components of the sun's total energy emission. One photon in the gamma ray or X-ray region has a large energy, but the sun emits only a small number of photons at these short wavelengths. Gamma rays are created in great abundance deep in the interior of the sun, but very few of them survive to escape into space. There is a rapid rise in emitted solar energy from the shortest wavelengths up to a maximum near 500 nm in the green part of the spectrum. The human eye evolved to respond to this most intense portion of the sun's emission between 400 and 780 nm. Beyond the visible region, the solar spectrum displays a long tail of energy that extends to wavelengths of several thousand nanometers, although beyond 2000 nm there is very little

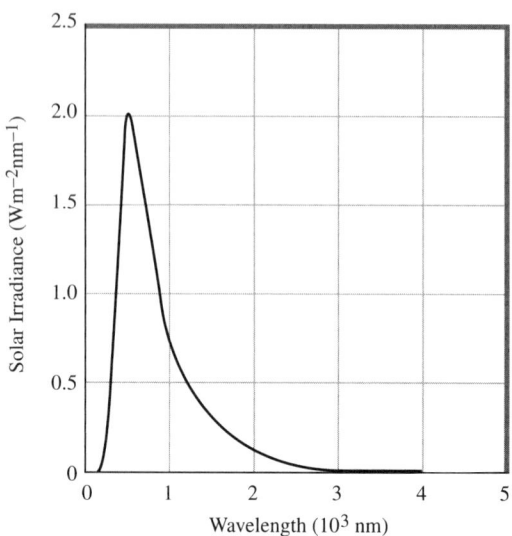

FIGURE 2.4   The extraterrestrial solar spectrum specifies the solar energy incident at the top of the Earth's atmosphere as a function of wavelength. Typical units are watts per square meter per nanometer of wavelength.

energy emitted. Table 2.2 illustrates how the sun's energy is distributed over several wavelength regions, as based on the compilation of Wehrli (1985) and the data of Neckel and Labs (1984).

Table 2.2 shows that the visible makes up only about half of the total energy emitted. The peak of the sun's spectrum lies in the green around 500 nm, but the entire visible emission of the sun consists of all colors mixed together. The human eye and brain perceive a mixture of colors as white. Based on this alone, the sun should be mainly white with a bias toward the green. Yet when the sun is high in the sky, the disk appears yellow. Why is this true? A large part of the answer involves scattering. As sunlight moves downward through the atmosphere, some of the energy is scattered out of the direct solar beam, and these photons move off into other directions. The direct beam that reaches the ground has been attenuated, and there is a sig-

TABLE 2.2   Distribution of Energy Emitted by the Sun

Spectral Region	Wavelength Range (nm)	% Total Energy Emitted
Ultraviolet and Shorter	0–400	~8%
Visible	400–780	~46%
Infrared and Longer	>780	~46%

nificant amount of scattered light moving in all directions. An elaborate mathematical theory describes the scattering of electromagnetic radiation by objects of various sizes (Bohren and Huffman 1983). In this case, the sunlight is being scattered by molecules of atmospheric nitrogen and oxygen, whose characteristic dimensions are small compared to the wavelength of light. The theory of scattering predicts that molecules scatter short-wavelength light much more efficiently than longer wavelengths, where the fraction of incident light that scatters in passing through a layer of air is inversely proportional to wavelength raised to the fourth power. Hence, short-wavelength visible sunlight, violet or blue, is scattered more efficiently than the longer wavelengths like red, and this is why the sky appears blue on a clear day. The diffuse blue glow coming from overhead is scattered sunlight. If the blue and violet light were put back into the solar beam, the sun would appear more white than yellow. Liou (2002) gives a more comprehensive discussion of atmospheric scattering processes.

Molecules are very small compared to the wavelengths of sunlight, and this fact is responsible for the wavelength dependence of molecular scattering. A different situation exists when the atmosphere contains liquid cloud droplets or ice crystals. The dimensions of these objects are usually larger than the wavelength of light. The theory of scattering shows that the efficiency of scattering by these "large" objects is almost independent of wavelength, so a cloudy sky scatters blue, yellow, and red light to almost the same degree. The result is that a cloudy sky looks white. A cloudy sky gives a truer representation of the color of the sun than does a clear sky, because after scattering by cloud droplets, all wavelengths are present in the diffuse light in almost the same proportion as they are in the sun's emission.

2.2 The Sun

The sun is a middle-aged star, around 4.6 to 4.8 billion years old. By earthly standards the sun is quite large. If the sun were a hollow sphere, approximately 1.3×10^6 Earths could fit inside it, and this size is typical for a star of the sun's class. Unlike many planets, however, the sun is not a solid object. The outer layers of the sun consist of hot gases, and the material becomes progressively denser and hotter as depth increases below the visible surface. Despite its considerable volume, the sun is not particularly massive. The total mass of the sun is only about 3.3×10^5 times that of the Earth, so the sun is made of relatively lightweight material. The mass density of liquid water is $\rho(H_2O) = 1.0 \, \mathrm{g \, cm}^{-3}$, and that of the solid Earth is $\rho(\mathrm{Earth}) = 5.5 \, \mathrm{g \, cm}^{-3}$. For comparison, the mass density of the sun averaged over its entire volume is $\rho(\mathrm{sun}) = 1.4 \, \mathrm{g \, cm}^{-3}$. The sun, taken as a whole, is only slightly denser than liquid water. The reason for the low solar density is that most of the sun's volume consists of hydrogen and helium, the two least massive atoms. These elements together make up approximately 98% of the sun's total mass (Gibson 1973).

The amount of mass present in the sun leads to very high pressures and temperatures in the core, where the local density is vastly greater than the average value given earlier. These extreme pressures and temperatures lead to generation of energy when atoms merge with each other to form heavier elements in a process called "nuclear fusion." Schematically, the simplest form of fusion is 4H → He + gamma rays. The hydrogen atoms taken collectively consist of four protons and four electrons, whereas the resulting helium atom has two protons, two neutrons, and two electrons. When fusion produces the helium atom, some energy is released, and it appears as kinetic energy in the helium and as a photon in the gamma ray part of the spectrum. Other fusion reactions proceed to build heavier elements. Nuclear fusion is confined to a small region in the core of the sun, which makes up about 1.5% to 2.0% of the solar volume. This is the only environment in which temperatures and pressures are sufficiently high to promote the merger of nuclei.

Virtually all of the radiant energy generated by the sun starts out as gamma rays in the core, but very little gamma radiation escapes into space. As gamma rays move outward through the solar interior, most of them are absorbed. This process heats the region of the sun located above the core, although these layers are cooler than those below. The visible surface of the sun has a temperature of 5700 to 5800 K, and material at this temperature emits a substantial fraction of its radiant energy in the visible and shorter wavelength infrared portions of the spectrum. The term "photosphere" refers to the visible surface of the sun, and the light that the human eye responds to arises mainly from this region. Still, nuclear fusion, creating gamma rays in the sun's core, is ultimately responsible for generation of visible light by the photosphere. In a very real sense, the existence of life on Earth depends on nuclear fusion occurring deep inside the sun.

2.3   The Solar Constant

Figure 2.4 illustrates the incoming solar energy as a function of wavelength when viewed above the atmosphere. However, for some purposes the relevant quantity is the total amount of solar energy received per unit area per unit time. This is the total area under the curve in Figure 2.4. Imagine a radiation detector, located above the Earth's atmosphere, pointed directly at the sun, so the solar beam strikes the sensor head-on. The detector responds to the total amount of solar energy striking a unit area in unit time, summed over all wavelengths present in sunlight. This quantity is the "solar constant," denoted by the symbol S_E. The definition is

$$S_E = \text{Total Solar Energy Crossing Unit Area in Unit Time}$$
$$\text{at the Top of the Earth's Atmosphere}$$

Implicit in this definition is the fact that the detector points straight at the sun. The currently accepted value, to three significant figures, is $S_E = 1.37 \times 10^3$ W m^{-2} (Willson 1997). A watt (W) has the dimensions of energy per unit time, where 1 W $= 1$ J s^{-1}. This is the same quantity used to label the energy per second emitted by a lightbulb. In general, a physical quantity whose dimensions are energy per unit area per unit time is called an "energy flux." Based on Table 2.2, approximately 8% of the solar constant is in the ultraviolet, 46% is in the visible, and 46% is in the infrared.

The Earth's orbit is slightly elliptical, so the distance to the sun varies over the course of a year. This changing distance leads to a systematic annual cycle in the amount of solar energy received. Some simple geometry makes this rigorous and identifies the quantities on which the solar constant depends. The sun emits energy into all directions in space. As this energy moves outward, it passes through spheres of progressively larger areas centered on the sun. Figure 2.5 illustrates the geometry. At a distance L from the center of the sun, the solar energy illuminates a sphere whose surface area is $4\pi L^2$. The total energy per second emitted by the sun is called the "solar luminosity," denoted by $\mathbb{E}$. The value of $\mathbb{E}$ ultimately derives from nuclear fusion reactions occurring in the sun's core, and a reasonable estimate is $\mathbb{E} = 3.86 \times 10^{26}$ W (Gibson 1973). Irrespective of the distance L, this same total amount of energy per second passes through any sphere centered on the sun.

The energy per second received by a unit area, $S(L)$, is simply the total energy per second crossing the sphere divided by its surface area:

$$S(L) = \mathbb{E}/(4\pi L^2) \qquad [2.3.1]$$

where the units are W m^{-2}. The length L can label the instantaneous distance of the Earth or any other planet from the sun. For Earth, L varies over a year

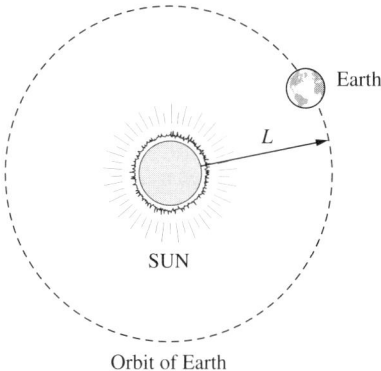

Orbit of Earth

FIGURE 2.5 The orbit of the Earth around the sun. Solar energy illuminates progressively larger spheres as it moves outward. The energy per second received by the planet depends on the solar energy output and the distance to the sun.

with the average $L_{ave} = 1.50 \times 10^{11}$ m. The solar energy flux received at the Earth for this average distance is

$$S(L_{ave}) = (3.86 \times 10^{26} \text{ W})/[4\pi(1.50 \times 10^{11}\text{m})^2] = 1.37 \times 10^3 \text{ W m}^{-2}$$

This value based on the annual average Earth–sun distance produces the solar constant $S_E = S(L_{ave})$. As a result of the elliptical character of the Earth's orbit, the solar energy flux in January, when the distance is a minimum, is about 6.9% larger than in July, when the planet reaches its farthest point from the sun.

The name "solar constant" is questionable terminology, because the solar luminosity $\mathbb{E}$ is not exactly constant over time. Considerable effort has focused on seeking variability in the solar constant. Historically, data obtained at high elevations were desirable since this minimizes the problem of confusing a change in the transmission of the atmosphere with a change in the sun's energy output. Today, satellite-based sensors perform measurements of the solar constant on an ongoing basis. During the two to three decades for which high-quality data are available, measured variations in S_E are quite small, on the order of $\pm0.1\%$ around the average value. Some of this change appears to represent a systematic trend over a decadal time scale (Willson 1997), and the rest arises from short-term variability associated with conditions on the sun that are akin to changes in weather on Earth. It is possible that measured changes in the solar constant largely reflect variability in ultraviolet wavelengths. For example, with reference to Table 2.2, a 1.2% to 1.3% change in the ultraviolet emission would cause an approximate 0.1% change in the solar constant. The 11-year solar cycle is the dominant mode of variation here. This cycle is most obvious in the number of sunspots on the solar disk, but these are accompanied by small changes in the solar luminosity as well. Lean and Rind (1998) present a more comprehensive review of the solar constant and its variability over time.

The variability of $\pm0.1\%$ refers only to the final quarter of the twentieth century, and this is an extremely short period in the life of the sun. Over the several-billion-year timescale of stellar evolution there surely have been major changes in the solar constant. Models of stellar evolution suggest that the present-day solar constant is significantly larger than what existed 3 to 4 billion years ago (Gough 1977). Differences as large as predicted would have led to a climate on the early Earth very different from what exists today. Problem 3 at the end of this chapter addresses this issue.

2.4 The Budget of Solar Radiation

The solar constant specifies the amount of energy received per unit area per unit time at the top of the atmosphere. However, the quantity of direct relevance to a theory of climate is the energy absorbed by the Earth and atmosphere, since this is what heats the planet. The "budget of solar radiation" defines, in a quantitative way, the possible fates of solar energy when it interacts with the atmosphere. The field of radiative transfer develops the theory required to describe how a beam of radiation is altered as it propagates through a medium. It is a mathematically rigorous subject, but the results are conceptually straightforward: some fraction of the incoming solar energy is absorbed by the atmosphere, another fraction is absorbed by the land and oceans, and, finally, the remaining fraction scatters back out the top of the atmosphere and escapes into outer space.

The development of a budget for solar radiation requires determining the fractions of incoming energy that go into each of these identified paths, where estimates of these quantities are derived from mathematical models that treat absorption and scattering of sunlight. Rather than detailing the calculations required, this discussion simply states some representative results. To set a scale, assume that 100 units of solar energy are incoming at the top of the atmosphere. This is a globally averaged number, related to the solar constant. Specifically, these 100 units represent the solar constant averaged over a time of 1 or more years as well as over the day and night sides of the Earth.

What happens to solar energy as it propagates downward? Some of it is absorbed in the atmosphere, and this totals 20 units, or 20% of the incoming energy (Houghton et al. 1996). Water and ozone are the two most important gaseous absorbers of solar radiation. Although ozone is known for its strong absorption in the ultraviolet part of the spectrum, this molecule also has a weaker absorption in the visible. Absorption by water vapor consumes a significant portion of the near-infrared solar radiation. In addition, liquid water and ice in clouds are responsible for a portion of the 20 units absorbed. Clouds cover approximately half of the Earth, and the liquid droplets and ice crystals of which they are composed are efficient reflectors of sunlight. This reflection is much more efficient than the absorption of solar radiation by clouds mentioned earlier. In addition to clouds, atmospheric molecules and particles reflect sunlight. All of these taken together scatter 22 units of solar energy back into space, where the contribution from clouds is the largest. To this point, 20 + 22 = 42 units of the original 100 units of solar energy have been accounted for.

The remaining 58 units reach the ground, but not all of this radiant energy is absorbed there. Some of the energy is reflected off the surface, and it eventually escapes into space. This portion is nine units, to which the polar ice caps make a significant contribution. Although ice covers a small fraction of the globe, it reflects a large percentage of the sunlight that reaches it. The ocean and land areas reflect only on the order of 10% of the solar energy that strikes them, where the exact value depends on details of the surface. This leaves 58 − 9 = 49 units of energy to be absorbed at the ground

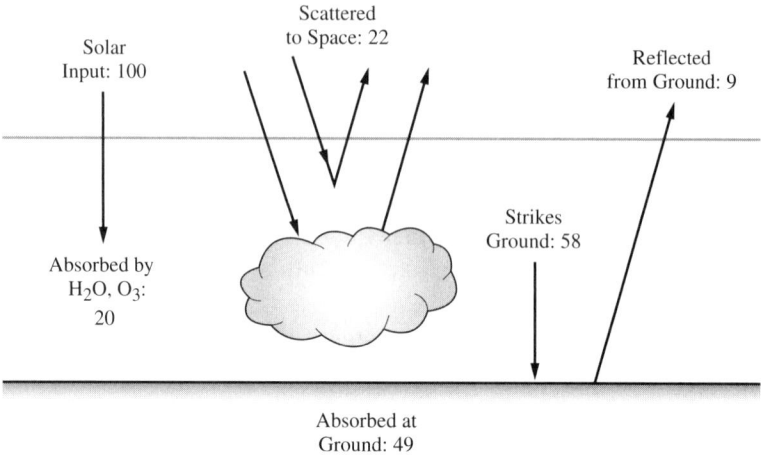

FIGURE 2.6   The budget of solar radiation in the Earth-plus-atmosphere system. Various percentages of the incident solar energy are scattered and absorbed by the atmosphere and the ground. [Values are based on fluxes reported in Houghton et al. (1996).]

(Houghton et al., 1996). Figure 2.6 illustrates the globally averaged budget of solar radiation in the Earth-plus-atmosphere system using the numbers stated above.

The "albedo" of the Earth-plus-atmosphere system, defined as the fraction of incoming solar energy reflected back to space, is an important parameter in the theory of climate. Figure 2.6 shows two contributions to the albedo, one from scattering in the atmosphere, 22 units, and one from scattering by the ground, 9 units. These values produce an albedo, A, of $A = (22 + 9)/100 = 0.31$, or to one significant figure 0.3. The largest single contributor to the albedo is reflection by clouds, but scattering by molecules and particles and reflection from the ground are significant as well. The albedo of the planet is not necessarily constant over time. For example, a change in the amount of particulate matter in the atmosphere would alter the amount of solar energy scattered back into space. Volcanic dust is a natural way to increase the Earth's albedo. The eruption of Mount Pinatubo in 1991 injected dust into the stratosphere, where it reflected sunlight for several years thereafter. The increased albedo promoted a slight drop in globally averaged temperature at the ground. The same mechanism operates on smaller spatial scales as well. Geographic areas where large quantities of smoke are released can be cooler during daytime than they would otherwise be because of the enhanced scattering of sunlight back to space.

A change in global cloudiness would alter the Earth's albedo, and it is feasible that human activity can alter cloud cover or cloud properties. Cur-

TABLE 2.3 The Budget of Solar Radiation

Absorption at Ground	49%
Absorption in Atmosphere	20%
Reflection to Space	31%
Total Solar Energy Input	100%

rent understanding (e.g., Houghton et al. 1996) indicates that a warming of the troposphere and the Earth's surface should occur over the next century in response to an increasing strength of the greenhouse effect (Section 2.7). One of the complications in predicting the magnitude of the temperature change centers on the response of global cloud cover to a warming. It is not necessarily true that cloudiness and the Earth's albedo will remain constant as the planet's temperature changes. The formation of clouds depends on evaporation of water and on vertical motions that transport water vapor up into the troposphere, where it condenses to form clouds. All these processes are dependent on the temperature at the ground, and their magnitudes likely will change as the Earth warms. A small increase in the Earth's albedo caused by increased cloudiness could promote a cooling that would counteract a portion of the greenhouse warming, whereas a decrease in cloudiness would act in the opposite direction.

Figure 2.6 presents the complete budget for solar radiation. In the end, there are only three things that can happen to incoming solar energy, and Table 2.3 lists these fates. The energy can be absorbed at the ground, and this component constitutes 49%. It can be absorbed in the atmosphere; this is 20%. Finally, the energy can be reflected back into space by clouds, gases, particles, and the ground, and this is a total of 31%. These three components sum, 49 + 20 + 31 = 100%, to equal the sun's energy input at the top of the atmosphere.

To this point some solar energy has been added to the Earth and atmosphere. This energy goes into heating the system, and in response the temperature rises. This is the first step in the sequence of processes that ultimately determines the Earth's equilibrium temperature. The following sections develop the concepts needed to examine the remaining parts of the energy balance.

2.5 Terrestrial Radiation

In addition to absorbing and scattering sunlight, the Earth and atmosphere emit radiation on their own, but this emission lies far into the infrared part of the spectrum, at wavelengths much longer than the sun's infrared radiation. The technical name for this radiant energy is "terrestrial radiation." To distinguish it from solar emission, the term "longwave radiation" is convenient, in which case sunlight takes the name "shortwave radiation." The balance of heating by absorption of shortwave radiation and cooling by emission of longwave radiation forms the foundation for a theory of planetary temperatures.

A classic derivation in thermodynamics shows that any solid or liquid object with an absolute temperature greater than zero emits electromagnetic radiation (e.g., Planck 1959). This general result applies in particular to the solid and liquid surface of a planet. In addition, certain gas-phase molecules emit longwave radiation, although this is not universally true. For example, the two most abundant gases in the Earth's atmosphere, N_2 and O_2, do not absorb or emit longwave radiation efficiently, a result that follows from details of their molecular structure. As a general result, a molecule constructed of two identical atoms (e.g., N_2, O_2, H_2) is not radiatively active in the longwave part of the spectrum. However, gas-phase molecules that consist of three or more atoms, commonly called "polyatomic molecules," have the potential to absorb and emit longwave radiation very efficiently. Prime examples of radiatively active gases in the Earth's atmosphere are CO_2, H_2O, and O_3.

Consider an object like the floor of a room or a thick layer of gaseous polyatomic molecules. How much longwave energy per second does the object or layer of gas emit? Specifically, how much energy comes from the surface of the object or the gaseous layer per unit area per unit time? The energy flux emitted by an object or by a layer of the atmosphere depends sensitively on its temperature, as does the way in which this radiant energy is distributed over wavelength. The "thermodynamic theory of radiation" predicts the energy flux that escapes from the surface of an object. An extremely important result derived from this body of reasoning, and confirmed by observations, is the "Stefan–Boltzmann Radiation Law." This is

$$\text{Energy Flux Emitted from a Surface} = \Phi = \sigma T^4 \qquad [2.5.1]$$

where T is absolute temperature, and the energy flux includes all of the photons emitted by the object, whatever their wavelengths may be. The energy flux is expressed in W m^{-2}, and σ is a universal constant called the Stefan–Boltzmann Radiation Constant, whose value is $\sigma = 5.67 \times 10^{-8}$ W m^{-2} K^{-4} (Hecht 1996). The emitting surface could be 1 m^2 of a solid object or the surface of the ocean. It could also be the surface of a thick layer of gas containing polyatomic molecules such as CO_2 and H_2O, although Section 2.8 identifies some complications in applying Eq. 2.5.1 to a gas. Notice that the chemical composition of the emitting substance does not appear in Eq. 2.5.1; all that matters is temperature. One could modify Eq. 2.5.1 to read $\Phi = \varepsilon \sigma T^4$, where ε in a fraction called the "emissivity" whose value can vary from one substance to another. Historically, an object for which $\varepsilon = 1.0$ is called a "black body," and this is a good approximation for the solid and liquid materials with which the atmosphere interacts. For a gaseous layer, however, ε might be substantially smaller than 1.0, depending on the chemical composition and thickness.

How much energy in the form of longwave radiation is emitted by 1 m^2 of a solid or liquid surface at a typical room temperature of 295 K, or about 72°F? Equation 2.5.1 says $\Phi = \sigma T^4 = (5.67 \times 10^{-8} \text{ W m}^{-2} \text{ K}^{-4}) (295 \text{ K})^4 = 429$ W m^{-2}. This is more energy per second from a square meter than seven 60-W lightbulbs, but at typical earthly temperatures this radiant energy lies in the far infrared part of the spectrum, where the human eye is unresponsive.

The Stefan–Boltzmann Radiation Law specifies the total radiant energy flux emitted by a surface, but it does not address how this energy is distributed over wavelength. Another result from the thermodynamic theory of radiation describes this wavelength dependence. This is "Planck's Radiation Law," named for Max Planck, who first derived the result in 1900 (Planck 1959). Planck's Radiation Law is a fairly complicated mathematical expression (see Eq. 2.5.2), but on a qualitative level what it says is quite simple and intuitive: hot objects preferentially emit high-energy photons, whereas colder objects preferentially emit lower-energy photons. Recall that the energy of a photon and the wavelength of light are inversely related. Hence, a hot object emits most of its radiant energy at short wavelengths, and a colder object emits at longer wavelengths. Of course, hot and cold are relative terms here.

The surface of the sun is hot, by earthly standards, and as a result it emits most of its energy at short wavelengths, in the visible part of the spectrum as well as in the shorter wavelength near-infrared. The energy flux emitted by the sun is characteristic of an object with a temperature of 5700 to 5800 K, with a peak emission in the green near 500 nm. On the other hand, a relatively cold object like the Earth emits radiation at much longer wavelengths, with a peak in the far-infrared near 10,000 nm. This longwave emission corresponds to a temperature of about 255 K, a result derived in Section 2.6. Figure 2.7 compares the wavelength dependence of the radiation emitted by two solid objects, one with a temperature similar to that of the sun's surface and the other with a temperature of 255 K. Note that the wavelength scale is logarithmic rather than linear. The behavior illustrates the appropriateness of the terms "shortwave" and "longwave" to describe the solar and terrestrial radiation fields, respectively. When applied to the Earth, Figure 2.7 is a simplification in that the observed emission of the planet-plus-atmosphere system has a wavelength dependence that is more complex than shown, and this arises from the presence of polyatomic gases in the atmosphere, a topic addressed in Section 2.8.

Planck's Radiation Law specifies the mathematical forms of the curves in Figure 2.7. Let this function be dB(λ,T)/dλ with the units W m^{-2} nm^{-1}. The derivation, done first by Planck and later by Einstein using a different approach (Van der Waerden 1967), yields

$$dB(\lambda,T)/d\lambda = (2 \times 10^{-9}) \pi (hc^2/\lambda^5) \{\exp[hc/(\lambda kT)] - 1\}^{-1} \quad [2.5.2]$$

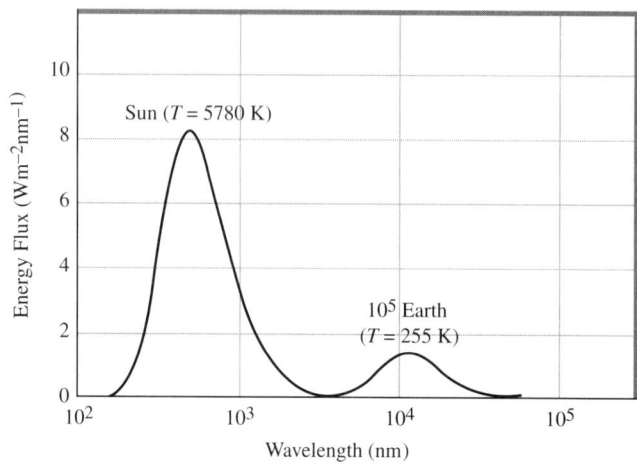

FIGURE 2.7 Thermal emission spectra of objects with temperatures similar to the sun (5780 K) and Earth (255 K). The solar surface emits 2.64×10^5 times as much energy per unit area per unit time as does the colder object.

where Planck's constant h is in J s, Boltzmann's constant k is in J K^{-1}, T is in K, and for numerical evaluation λ is in m. When $T = 5700$–5800K, the values of $dB(\lambda,T)/d\lambda$ are similar to the magnitude and wavelength dependence of sunlight. When $T = 255$K, this same function produces a wavelength-dependent energy flux that approximates the longwave energy emitted by a solid object whose temperature is like that of the Earth-plus-atmosphere system. Planck's expression and the Stefan–Boltzmann Radiation Law are related by

$$\Phi = \sigma \, T^4 = \int_0^\infty d\lambda \, dB(\lambda,T)/d\lambda \qquad [2.5.3]$$

When the energy flux per unit wavelength, $dB(\lambda,T)/d\lambda$, is integrated over all wavelengths, the result is the total energy flux, σT^4. The thermal emission spectrum representing the Earth in Figure 2.7 has a wavelength dependence similar in shape to that of the sun, but there are two important differences. First, the terrestrial emission is shifted to much longer wavelengths than that of the sun, as predicted by Planck's Radiation Law. Second, the total energy flux lost by a unit area of the Earth-plus-atmosphere system is vastly smaller than that for the sun. To quantify this, assume the temperature that describes the sun's emission to be 5780 K, and for the Earth adopt 255 K. These numbers, together with the Stefan–Boltzmann Radiation Law, imply that each square meter of the sun's surface emits $(5780/255)^4 = 2.64 \times 10^5$ times the energy lost by a unit area of the Earth as measured at the top of the atmosphere.

A fundamental question is, Why should an object or a layer of polyatomic gas emit electromagnetic radiation simply because its temperature is greater than absolute zero? What is taking place at the molecular level to create this radiant energy? The situation is easiest to visualize by considering a volume of gas, some fraction of which is carbon dioxide at a known temperature. Because the temperature is greater than absolute zero, the molecules are moving about at random and colliding with each other. As described in Section 1.4, molecules have both kinetic and internal energy. Carbon dioxide is a linear molecule, meaning that the carbon atom and the two oxygen atoms are arranged in a straight line, as O-C-O, but the atoms can vibrate, stretch, and bend around these positions. These modes of motion constitute internal energy. For example, the C atom can move up while both of the O atoms move down, followed by motions in the opposite directions. This is called a bending mode, in which the CO_2 molecule is somewhat like a bird flapping its wings. In another mode, the atoms can remain stationary with respect to each other. The bending motions of the atoms in the molecule constitute an "excited state," and the "ground state" corresponds to the three atoms remaining in fixed positions.

A CO_2 molecule in its ground state moves about through the gas, and as a result of a collision, it can gain some internal energy and enter an excited state. This process, called "collisional excitation," is written as

$$CO_2 + M \rightarrow CO_2{}^* + M \qquad [2.5.4]$$

The * indicates that the CO_2 is an excited state. The molecule M, which in the Earth's atmosphere is most likely N_2 or O_2, loses kinetic energy in the collision, where the amount corresponds to the spacing between the two states in the CO_2 molecule. Process 2.5.4 can occur only if the molecule M has at least this minimum amount of kinetic energy. The excited $CO_2{}^*$ can give up its extra energy in two ways. One is via another collision:

$$CO_2{}^* + M \rightarrow CO_2 + M \qquad [2.5.5]$$

Process 2.5.5 is called "collisional deactivation." Here the $CO_2{}^*$ loses internal energy, and the molecule M gains an equal amount of kinetic energy. An alternate path is for the $CO_2{}^*$ to give up its energy by emitting a photon in a process called "thermal emission":

$$CO_2{}^* \rightarrow CO_2 + photon \qquad [2.5.6]$$

Energy that was contained in the excited state is lost via creation of a photon. The name "thermal radiation" is appropriate because this energy started out as kinetic energy, which is measured by temperature. If the carbon dioxide molecule is in the Earth's atmosphere, the photon could escape into outer space, resulting in a loss of energy from the planet. In this case, energy that began as kinetic energy of the molecule M in process 2.5.4 ends up in a photon that moves into outer space as a result of process 2.5.6. Energy escaping from an object or from the top of the atmosphere is a cooling effect since energy is lost from the system.

This process also can work in reverse. Suppose a radiation field is imposed on a gas from outside. For example, the surface of the Earth emits longwave radiation up into the atmosphere, and molecules of CO_2 absorb some of these photons:

$$CO_2 + photon \rightarrow CO_2^* \qquad [2.5.7]$$

The excited molecule may give up its energy in a collision, process 2.5.5, or it may radiate in process 2.5.6. If process 2.5.5 occurs, energy goes into heating the atmosphere. However, if the photon escapes into space, the result is a cooling of the system. This example uses carbon dioxide, but water vapor, ozone, and other polyatomic molecules behave the same way. As noted previously, N_2 and O_2 are not radiatively active in the longwave part of the spectrum because of details of their molecular structure. The most important molecules for absorption and emission of longwave radiation in the Earth's atmosphere are not the major gases but instead are polyatomic trace gases.

2.6   The Radiative Energy Balance

The combination of solar (shortwave) and terrestrial (longwave) radiation leads to the concept of "radiative equilibrium." The Earth continuously receives energy from the sun, and part of this energy is absorbed, thereby heating the planet. If heating by sunlight were the only process at work, the temperature of the Earth would have become progressively warmer over time. Of course, this has not happened, and the explanation is simple. Over a given period, the Earth and atmosphere combined must emit as much energy back into space as they absorb from the sun, corresponding to the balanced state of radiative equilibrium. This is the only way the planet can maintain a constant temperature. The mathematical statement of radiative equilibrium is

[Total Solar Energy Absorbed per Second]
    = [Total Terrestrial Energy Lost to Space per Second]   [2.6.1]

If the two quantities in the equation were not equal, then the Earth and atmosphere would be either heating up or cooling down over time. This balance is also expressible in terms of heating and cooling rates:

[Heating Rate from Absorption of Solar Radiation]
    = [Cooling Rate from Loss of Terrestrial Radiation]   [2.6.2]

Absorption of sunlight heats the Earth and atmosphere, and loss of long-wave radiant energy to outer space effectively cools the system. The balance expressed by Eqs. 2.6.1 and 2.6.2 applies when averaged over the entire planet for at least a full year.

The equality of heating and cooling rates is not an accident; the Earth-plus-atmosphere system naturally will come into a state of radiative equilibrium. Imagine the planet starts out with a temperature of absolute zero, and there is no sunlight. If $T = 0$ K, the Stefan–Boltzmann Radiation Law, Eq. 2.5.1, says that no terrestrial radiation is emitted. Now suppose the sun turns on, and solar energy strikes the Earth. Some fraction of this energy is absorbed and heats the planet, so the temperature rises above 0 K. When $T > 0$ K, the planet starts to emit thermal radiation back into space, as described by Eq. 2.5.1. Each square meter of the surface emits an energy flux equal to σT^4, and this loss of energy acts to cool the planet. At this stage, heating from absorption of shortwave sunlight competes with cooling by loss of longwave radiation. The hotter the planet becomes from absorbing sunlight, the more longwave radiation it loses, and this counteracts the heating. Eventually the planet warms up to an average temperature where, over a period of 1 year, it emits as much radiant energy into space as it absorbs from the sun. Once this balance is reached, the planet is neither heating up nor cooling down, and the temperature stops changing with time. This is the state of radiative equilibrium. In radiative equilibrium, the shortwave energy absorbed from the sun, averaged over the entire Earth for a full year, equals the energy lost to space as longwave terrestrial radiation.

It is possible to convert this qualitative explanation into a mathematical model to compute the temperature of the Earth or any other planet. Let E(absorbed) be the total solar energy per second absorbed by the Earth-plus-atmosphere system, and let E(lost) be the total terrestrial energy per second that escapes from the top of the atmosphere into outer space. Both E(absorbed) and E(lost) are expressible in watts. Radiative equilibrium says

$$E(\text{absorbed}) = E(\text{lost}) \qquad [2.6.3]$$

Consider E(lost) first. If the Earth is a sphere of radius R, its surface area is $4\pi R^2$. Strictly, R should be the radius of the planet plus the depth of the atmosphere responsible for emitting most of the longwave radiation into space; however, this is negligibly different from the radius of the solid Earth. The Stefan–Boltzmann Radiation Law specifies the energy lost per unit area per unit time. Let this energy flux be σT_E^4, where T_E is the "effective radiating temperature" of the Earth-plus-atmosphere system. The longwave energy lost to space is

$$E(\text{lost}) = 4\pi R^2 \, \sigma \, T_E^4 \qquad [2.6.4]$$

Exactly what does T_E refer to? The temperature of the Earth's surface varies with latitude, and the temperature of the atmosphere varies in both altitude and latitude. For the moment, the best approach is to interpret the effective radiating temperature as some form of average over both the Earth and the part of the atmosphere responsible for the loss of longwave radiation to space. A more precise definition is forthcoming.

The other piece of radiative equilibrium involves the solar energy absorbed per second. Some care is needed to define the appropriate geometry because solar energy approaches the Earth from only one direction in space. The solar constant S_E specifies the energy that crosses unit area in unit time at the top of the atmosphere, where the reference area is oriented perpendicular to the flow of radiation. The solar energy that strikes the planet passes through a circle of radius R, with area πR^2, so the total energy incident per second is $\pi R^2 S_E$. The relevant quantity for radiative equilibrium is the energy per second absorbed by the Earth-plus-atmosphere system. Some of the incident energy is reflected back to space, and the albedo defined in Section 2.4 specifies this fraction. If A is the albedo, then $1 - A$ is the fraction of incoming energy that is absorbed. Given this, the total solar energy per second absorbed by the Earth-plus-atmosphere system is

$$E(\text{absorbed}) = (1 - A)\pi R^2 S_E \qquad [2.6.5]$$

The mathematical statement of radiative equilibrium based on Eqs. 2.6.3, 2.6.4, and 2.6.5 is

$$(1 - A)\pi R^2 S_E = 4\pi R^2 \sigma T_E^4 \qquad [2.6.6]$$

where Figure 2.8 illustrates the geometry. Eq. 2.6.6 constitutes a definition of "effective radiating temperature." Specifically, the effective radiating temperature provides just enough longwave energy escaping to space to balance the shortwave solar energy that is absorbed by the Earth-plus-atmosphere system under the assumption that the Stefan–Boltzmann Radiation Law describes the longwave emission. Another way to think about Eq. 2.6.6 is in terms of heating and cooling rates. The effective radiating temperature provides just enough cooling to space to balance the heating from absorption of sunlight.

Eq. 2.6.6 shows that the average temperature of a planet assumes the value required to balance the shortwave heating rate. If a change took place in the solar constant or in the albedo of the system, the amount of energy absorbed per second would change. In response to this, the effective radiating temperature would adjust just enough to restore radiative equilibrium. The effective radiating temperature, based on Eq. 2.6.6, is

$$T_E = [(1 - A) S_E/(4\sigma)]^{1/4} \qquad [2.6.7]$$

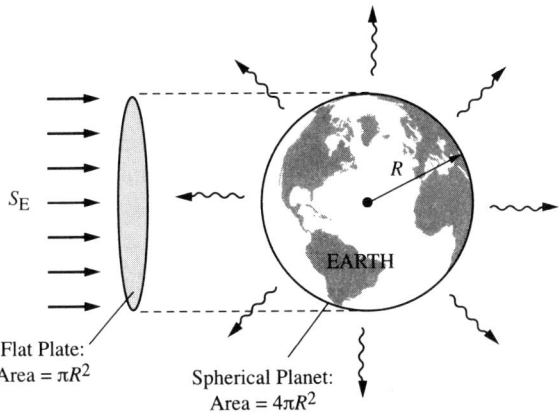

S_E

Flat Plate:
Area = πR^2

Spherical Planet:
Area = $4\pi R^2$

FIGURE 2.8   Radiation balance of the Earth-plus-atmosphere system. The total solar energy absorbed during 1 year is balanced by the terrestrial energy lost to space.

where the radius of the planet cancels from the final expression. Numerical values appropriate for Earth are $A = 0.3$ and $S_E = 1.37 \times 10^3$ W m^{-2}, and the Stefan–Boltzmann Radiation Constant is $\sigma = 5.67 \times 10^{-8}$ W m^{-2} K^{-4}. These produce $T_E = 255$ K for Earth, which on the Fahrenheit scale lies between $0°$ and $-1°$F.

Temperatures as low as the value just computed certainly exist at some locations on the surface of the Earth, but the globally averaged temperature at the ground is considerably warmer, being $T_G = 288$ K or about $59°$F. In view of the low value for T_E, what does the effective radiating temperature actually measure? A key concept here is that radiative equilibrium is a condition that applies at the top of the Earth's atmosphere. It concerns the incoming solar energy and the longwave terrestrial energy lost to outer space. Most of this longwave radiation originates in the troposphere, a region where the average temperature is colder than at the ground. Given this, one would expect the effective radiating temperature to be less than the temperature at the Earth's surface. Roughly half of the total mass of the atmosphere lies above an altitude of 5 km and approximately half is below, and the temperature at 5 km up is very near 255K (*U.S. Standard Atmosphere, 1976*). The effective radiating temperature is indicative of the atmosphere's temperature averaged over the altitude region that produces the longwave energy lost to space.

2.7   The Greenhouse Effect: A Simplified Model

The problem remains of explaining why the surface temperature is higher than the effective radiating temperature, and this leads to the "greenhouse effect." If the Earth had no atmosphere, that is, if it were a bare rock like the moon, then the effective radiating temperature would be the same as the surface temperature averaged over the entire globe for periods of 1 or more years. However, when an atmosphere containing radiatively active polyatomic molecules surrounds a planet, the situation becomes more complex. The observed averaged temperature at the ground is $T_G = 288$ K, warmer than the computed effective radiating temperature by 33 K, or 59° to 60°F. According to the Stefan–Boltzmann Radiation Law, each square meter of the Earth's surface emits radiant energy up into the atmosphere. The energy flux is related to the surface temperature via

$$\Phi(\text{ground}) = \sigma \, T_G{}^4 = (5.67 \times 10^{-8} \text{ W m}^{-2} \text{ K}^{-4})(288 \text{ K})^4 = 390 \text{ W m}^{-2}$$

The observed surface temperature requires a globally averaged energy flux of 390 W m^{-2} to flow upward from the ground into the atmosphere. However, the longwave energy flux that escapes into outer space is described by the effective radiating temperature of $T_E = 255$ K. This is

$$\Phi(\text{lost}) = \sigma \, T_E{}^4 = (5.67 \times 10^{-8} \text{ W m}^{-2} \text{ K}^{-4}) \, (255 \text{ K})^4 = 240 \text{ W m}^{-2}$$

When summed over the entire planet, this 240 W m^{-2} maintains radiative equilibrium. Obviously, some energy is vanishing somewhere. The ground emits 390 W m^{-2}, but only 240 W m^{-2} escape to space, leaving $390 - 240 = 150$ W m^{-2} unaccounted for. The missing ingredients here are atmospheric polyatomic gases that absorb much of the longwave radiation emitted by the ground. The two most important longwave absorbers are CO_2 and H_2O, but O_3, methane (CH_4), and the manmade chlorofluorocarbons are also significant. Because of this absorption, most of the longwave radiation that originates at the ground is absorbed in the atmosphere.

Absorption by polyatomic molecules is only part of the greenhouse effect; absorption of longwave radiation is followed by emission. Figure 2.9 illustrates the missing information. Some of the reemitted energy goes upward, escapes into space, and supplies the 240 W m^{-2} needed to maintain radiative equilibrium. The remainder of the reemitted energy goes downward toward the ground, where it is absorbed. The longwave radiation coming from the atmosphere warms the ground in the same way as sunlight, and this extra energy makes the surface of the Earth warmer than it would otherwise be.

It is possible to derive a mathematical expression to show how the greenhouse effect increases the surface temperature of a planet. This particular example, patterned after the discussion in Goody and Walker (1972), divides the atmosphere into three layers, although the same reasoning applies to an arbitrary number of layers. Let the average atmospheric temperature in each layer, going from high altitudes to low, be T_1, T_2, and T_3, respectively,

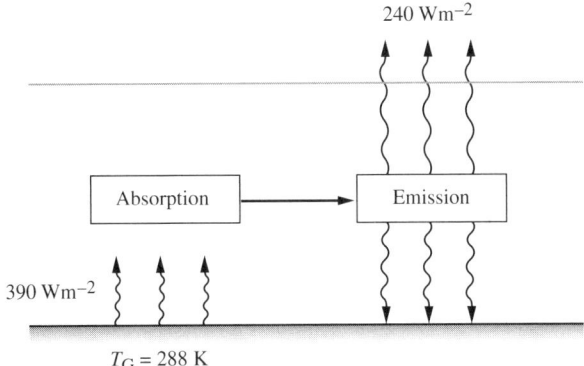

240 Wm^{-2}

Absorption → Emission

390 Wm^{-2}

$T_G = 288$ K

FIGURE 2.9   The greenhouse effect. Polyatomic molecules in the atmosphere absorb longwave energy emitted by the ground. The molecules then radiate this energy both upward and downward. The downwelling longwave energy heats the ground.

and the temperature of the ground is T_G, as shown in Figure 2.10. From radiative equilibrium, the Earth and atmosphere taken together absorb a shortwave energy per second equal to $\pi R^2(1 - A)S_E$. This example assumes that all of this solar energy is absorbed at the ground, although this is a simplification since the budget of solar radiation shows that the atmosphere absorbs 20% of the incoming shortwave energy. Problem 4 at the end of this chapter considers how atmospheric absorption of solar energy modifies the following derivation.

In radiative equilibrium, the total shortwave energy absorbed by the planet and atmosphere combined is balanced by the longwave energy lost to space. It is convenient to think in terms of the globally averaged solar energy absorbed per unit area of the planet per second. This is just the total solar energy absorbed per second divided by the surface area of the planet. From Eq. 2.6.6 this is

$$\pi R^2(1 - A)S_E/(4\pi R^2) = \sigma\, T_E^{\,4} \qquad [2.7.1]$$

Radiative equilibrium requires that the average solar energy absorbed per unit area of the planet per second equal $\sigma T_E^{\,4}$, and by assumption, all of this energy is absorbed at the ground.

Each of the three layers in Figure 2.10 contains polyatomic molecules that both absorb and emit longwave radiation. For ease of discussion, this derivation assumes that CO_2 is the only radiatively active gas, although the true atmosphere contains a mixture of several greenhouse gases as well as clouds, which behave in a way similar to gases. In this idealized example, it is necessary to choose the thicknesses of the layers so that each one contains the same number of CO_2 molecules. The thickness of each layer is defined so there is just enough CO_2 to absorb 100% of the longwave radiation that

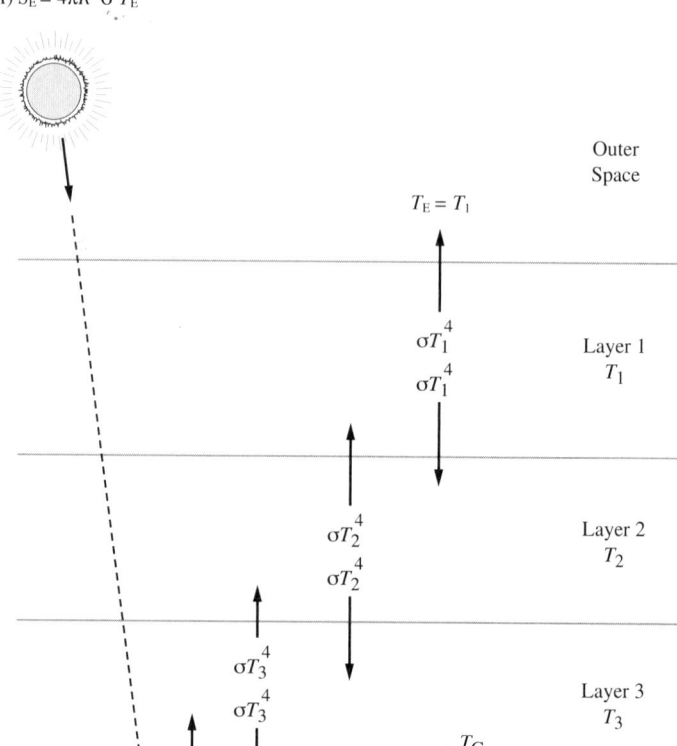

$\pi R^2 (1-A) S_E = 4\pi R^2 \sigma T_E^4$

FIGURE 2.10 A simplified model of the greenhouse effect. Solar energy is absorbed at the ground. Each atmospheric layer emits the same amount of longwave energy as it absorbs from adjacent layers.

comes from the adjacent layer, both above and below. There is no more and no less CO_2 than this special amount. For example, layer 3 is next to the planetary surface, and each unit area of the ground emits an energy flux equal to σT_G^4 upward into layer 3. The thickness of layer 3 is defined to be just enough so that the CO_2 in it absorbs 100% of the longwave energy coming up from the ground and 100% of the longwave energy coming down from layer 2. Each of the layers has this same CO_2 abundance, called "one extinction thickness." The term "one extinction thickness" as used here means that there is just enough CO_2 in a layer to absorb 100% of the incident longwave energy coming from adjacent layers.

Each of the three layers in Figure 2.10 emits radiant energy according to its temperature as described by the Stefan–Boltzmann Radiation Law, with the same energy flux going both upward and downward. All of the longwave radiation that escapes into outer space has to originate in layer 1, since the specially chosen thickness of the layer requires this to be true. Based on the definition of effective radiating temperature, T_1 must be the same as T_E, so the first conclusion is

$$T_1 = T_E \qquad [2.7.2]$$

where T_E is known from Eq. 2.6.7. The effective radiating temperature describes the energy lost to space, and all of this energy comes from layer 1.

For temperature to remain constant in time, each layer and the ground have to lose as much energy per unit time as they absorb:

[Energy Flux Absorbed by Layer i] = [Energy Flux Lost by Layer i]
$$[2.7.3]$$

where i identifies an atmospheric layer or the ground. This requirement, combined with the Stefan–Boltzmann Radiation Law, determines the temperature of each layer and the ground. Beginning at the top, layer 1 absorbs all of the energy coming up from layer 2, a flux equal to $\sigma\, T_2^4$. Then layer 1 emits both upward and downward, where the same flux, $\sigma\, T_1^4$, applies to both directions. The balance is

$$\sigma\, T_2^4 = \sigma\, T_1^4 + \sigma\, T_1^4 \qquad [2.7.4]$$

Since $T_1 = T_E$, the temperature of layer 2 satisfies $T_2^4 = 2\, T_E^4$ or

$$T_2 = (2)^{1/4}\, T_E \qquad [2.7.5]$$

Next, application of Eq. 2.7.3 to layer 2 gives

$$\sigma\, T_1^4 + \sigma\, T_3^4 = \sigma\, T_2^4 + \sigma\, T_2^4 \qquad [2.7.6]$$

The energy absorbed in layer 2 comes down from layer 1 and up from layer 3. Since both T_1 and T_2 have been expressed in terms of the known T_E, Eq. 2.7.6 produces a result for T_3. This is $T_3^4 = 2\, T_2^4 - T_1^4 = 2\,(2T_E^4) - T_E^4 = 3\, T_E^4$ or

$$T_3 = (3)^{1/4}\, T_E \qquad [2.7.7]$$

Finally, the energy balance in layer 3 determines the temperature of the ground. Layer 3 absorbs energy emitted by layer 2 and by the ground. The energy balance from Eq. 2.7.3 is

$$\sigma\, T_2^4 + \sigma\, T_G^4 = \sigma\, T_3^4 + \sigma\, T_3^4 \qquad [2.7.8]$$

The temperature of the ground satisfies $T_G{}^4 = 2T_3{}^4 - T_2{}^4 = 2(3\ T_E{}^4) - 2T_E{}^4 = 4T_E{}^4$ or

$$T_G = (4)^{1/4}\ T_E \qquad\qquad [2.7.9]$$

An analysis of the energy balance of the ground also produces Eq. 2.7.9. In this case, the energy absorbed is both shortwave, $\sigma T_E{}^4$ via Eq. 2.7.1, and longwave, from layer 3, where $\sigma T_3{}^4 = 3\sigma T_E{}^4$. The emission of the ground is into the upward direction only, so Eq. 2.7.3 applied to the planetary surface instead of to an atmospheric layer is

$$\sigma\ T_E{}^4 + \sigma\ T_3{}^4 = \sigma\ T_G{}^4 \qquad\qquad [2.7.10]$$

and leads to Eq. 2.7.9. The solutions for T_1, T_2, T_3, and T_G in terms of the known effective radiating temperature T_E, computed in Eq. 2.6.7, constitute the "radiative equilibrium temperature profile."

This model produces a downward flux of longwave energy at the ground that is greater than the flux of solar shortwave energy. The downward flux of solar radiation, from Eq. 2.7.1, is equal to $\sigma T_E{}^4$, while the greenhouse model for a three-layer atmosphere produces a longwave flux at the ground equal to $\sigma T_3{}^4 = 3\sigma T_E{}^4$. In this particular model, the atmospheric longwave energy at the ground is three times larger than the solar shortwave energy. This is how the greenhouse effect warms the surface of a planet; longwave radiation from the atmosphere warms the ground in the same way as solar radiation. An inequality, where the longwave energy exceeds the shortwave contribution, is indeed the case on Earth, although the excess of longwave over shortwave is not as large as predicted by the three-layer model.

The preceding example assumes an atmosphere with three extinction thicknesses of the greenhouse absorbing gas. The same reasoning, based on repeated use of Eq. 2.7.3, applies to any atmospheric extinction thickness, so long as it is an integer. In general, the temperature of the ith layer measured from the top of the atmosphere is

$$T_i = (i)^{1/4}\ T_E \qquad\qquad [2.7.11]$$

and the temperature at the ground is

$$T_G = (1 + \tau)^{1/4}\ T_E \qquad\qquad [2.7.12]$$

where τ is the total number of extinction thicknesses of the longwave absorber, also called the "total extinction thickness."

Because of the greenhouse effect, the surface of the Earth is about 33 K, or 59° to 60°F, warmer than it would otherwise be. This natural greenhouse effect makes the difference between an Earth where liquid water can exist and a planet where much of the water would be frozen. The greenhouse effect often appears in the context of global warming. Widespread burning of fossil fuels is adding CO_2 to the atmosphere, and this is increasing the extinction thickness in the longwave part of the spectrum. The result should be a warming of the planet. Notice that an increase in the abundance of atmospheric CO_2 does not influence the effective radiating temperature computed from Eq. 2.6.7. Instead, the effect is to increase τ in Eq. 2.7.12. In this simplified model, the flux of shortwave solar energy absorbed and the longwave energy escaping to space at the top of the atmosphere remain unchanged as the surface temperature increases.

On Earth, the greenhouse gases are trace constituents, but what would happen on a planet where a greenhouse gas is a major component of the atmosphere? Earth's nearest neighbor, Venus, has an atmosphere that contains 96% to 97% CO_2 by volume. Furthermore, the surface pressure is 90 times that of Earth. For this and other reasons identified later, the greenhouse effect on Venus is very strong and leads to a surface temperature in the vicinity of 750 K (Christiansen and Hamblin 1995).

2.8 The Atmospheric Energy Balance: Some Complications

The radiative equilibrium temperature profile derived in the preceding assumes that exchange of longwave radiation between adjacent layers, acting alone, determines the temperature structure of the atmosphere, whereas the temperature of the ground involves both shortwave and longwave radiation. This model successfully illustrates the mechanism of the greenhouse effect, but it fails to capture the complexity of the real atmosphere. A major simplification in Section 2.7 is the assumption that energy moves between the Earth's surface and the atmosphere only in the form of radiation. The pure radiative treatment predicts a discontinuity in temperature between the ground and the atmosphere adjacent to it. This temperature difference gives rise to thermal convection, identified in Section 1.5, and these motions transport heat from the surface into the atmosphere. Finally, evaporation of water acts to cool the ground, while condensation back to liquid or ice warms the atmosphere, as described in Chapter 3. As a consequence, the actual temperature profile, discussed in Section 1.4, differs from the prediction of radiative equilibrium.

Irrespective of energy transport by means other than radiation, the model developed in Section 2.7 is based on an assumption involving the longwave extinction thickness. The thickness of each atmospheric layer is defined so that it absorbs 100% of the longwave energy incident from above or below. The problem is that it is not possible to achieve this complete absorption in practice. To see why, one must consider the structure of gas phase polyatomic molecules. The surface of the Earth emits longwave energy spread over a broad range of wavelengths upward into the atmosphere. The total energy flux is σT_G^4, and Planck's Radiation Law describes the distribution of this radiant energy as a function of wavelength. If there were no atmosphere, all of this energy would escape into space, but instead the spectrum of radiation emitted by the ground passes through an atmosphere containing molecules such as CO_2 and H_2O. The important point, neglected in Section 2.7, is that polyatomic molecules do not absorb radiation with equal efficiency at all of the wavelengths emitted by the ground. For example, CO_2 absorbs and emits radiation at wavelengths around 15,000 nm (15 μm), but it has little or no effect on radiation at other wavelengths, such as 10,000 or 30,000 nm. This traces back to the energy level structure of a CO_2 molecule. As described in Section 2.1, a free molecule will absorb and emit only those photons whose energies match the spacing between two energy levels.

In a solid or liquid, the discrete energy levels that exist in the gas phase are effectively broadened by interactions between the tightly packed molecules. Hence, the surface of the Earth emits a smooth spectrum of radiation similar to that in Figure 2.7 but described by a globally averaged temperature of 288 K. As this spectrum propagates upward, CO_2 selectively absorbs photons in a band of wavelength around 15,000 nm. The absorption and re-emission are not confined to a single wavelength because one CO_2 molecule contains many different energy levels that correspond to numerous wavelengths near 15,000 nm. In addition, CO_2 molecules in the atmosphere experience a large number of collisions each second. In these collisions, energy levels are perturbed with the result that a range of wavelengths in the vicinity of 15,000 nm may be absorbed and emitted. A layer in the atmosphere could contain just enough CO_2 to absorb 100% of the incident radiation in a narrow range of wavelength around 15,000 nm, but this same layer would be almost transparent to radiation at wavelengths further removed from 15,000 nm.

The model of the greenhouse effect in Eq. 2.7.12 implicitly assumes that polyatomic molecules absorb and emit at all wavelengths across the longwave spectrum, but in fact that is not the case. Molecules in the gas phase interact only with photons in certain wavelength ranges. For example, ozone has a strong absorption feature near 9600 nm in wavelength, close to the peak of the solid Earth's longwave emission, while water vapor has numerous absorption features throughout the longwave spectrum. Consequently, the spectrum of longwave radiation that escapes from the top of the atmosphere into space is not a smooth function of wavelength as depicted in Figure 2.7. Instead, satellite-based measurements, available since the early 1970s (Hanel et al. 1972), reveal a wavelength-dependent structure like that shown in Figure 2.11. Ozone is responsible for a dip in the spectrum around 9600 nm, and CO_2 causes a deep minimum near 15,000 nm. The less-pronounced jagged structure spread over the spectrum arises from water va-

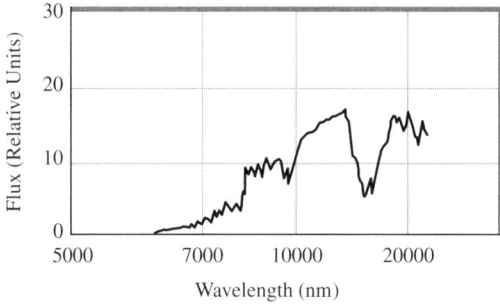

FIGURE 2.11   The longwave spectral flux of radiation lost to space at the top of the Earth's atmosphere. Structure in the spectrum arises from absorption by polyatomic atmospheric molecules. [Based on original data reported by Hanel et al. 1972.]

por. The total energy flux lost to space is equal to σT_E^4, but the wavelength dependence of this total energy flux is complicated. It is not a smooth curve, as one would predict from a Planck function evaluated at a fixed temperature.

Eq. 2.7.12 illustrates the physical mechanism of the greenhouse effect, but it cannot account for complications that arise from wavelength-dependent absorption. According to the simple model, if the total extinction thickness τ approaches infinity, then the temperature of the ground must approach infinity as well. Fortunately, this is not possible from an increase in the atmospheric CO_2 abundance alone. This is because a single type of gas phase polyatomic molecule is incapable of absorbing all wavelengths of the longwave radiation emitted by the ground.

For the surface temperature to increase dramatically, the atmosphere would need to become an efficient absorber across the entire longwave spectrum. While this appears unlikely on Earth, there is a planet in the solar system that approaches this extreme condition. Section 2.7 mentioned the greenhouse effect on Venus. Even in the dense atmosphere of this planet, CO_2 acting alone would not produce a surface temperature in excess of 700 K. However, Venus is totally enveloped by clouds, and their presence has a major influence on the surface temperature. When tightly packed in the liquid phase, molecules are able to absorb and emit over the entire longwave spectrum because the energy levels are broadened by strong interactions. Eq. 2.7.12 describes the longwave trapping by a cloud quite well, and the efficient greenhouse effect provided by thick cloudiness is a major contributor to the high temperature on Venus.

The greenhouse effect of clouds on Earth is well known, and it is most apparent at night when longwave cooling does not compete with solar heating. A cloudy night tends to have warmer temperatures than a clear night at the same time of the year. The clouds absorb the longwave radiation emitted by the ground over a broad range of wavelength, and they reemit some of it back downward. This extra energy coming from the clouds is a source of heating, and it slows the nocturnal cooling of the ground. Clouds have both a cooling effect by increasing the planet's albedo during daylight and a warming effect by increasing longwave optical thickness. For a specific cloud, it is not possible to tell which effect dominates without an analysis of the associated optical properties and temperature.

2.9   Latitudinal Dependence of Heating and Cooling: The Seasonal Cycle

The concept of radiative equilibrium is central to the theory of climate, but it applies only in a globally and annually integrated sense. The total amount of solar energy per second absorbed is $\pi R^2(1 - A)S_E$, where this is a sum over the entire planet averaged for a year or more. Similarly, $4\pi R^2\sigma T_E^4$ is the globally integrated terrestrial energy lost to space. A complete theory should also consider the solar energy absorbed (heating) and the terrestrial energy lost (cooling) as functions of latitude and time of year. Figure 2.12 depicts these quantities for dates near March 21 and September 21 based on satellite data (Vonder Haar and Suomi 1971). On these dates the sun is directly overhead at noon as viewed from the equator. The general latitudinal pattern in Figure 2.12 is a simple result of the Earth's spherical geometry. A beam of sunlight incident over the equator spreads over a smaller area projected onto the Earth's horizontal surface than an identical beam of sunlight approaching at a higher latitude.

Consider the beam of light striking the high-latitude region in Figure 2.13, and let S_E be the energy flux moving perpendicular to the area A. Based on simple geometry, the flux that strikes the horizontal area a is $(A/a)S_E =$

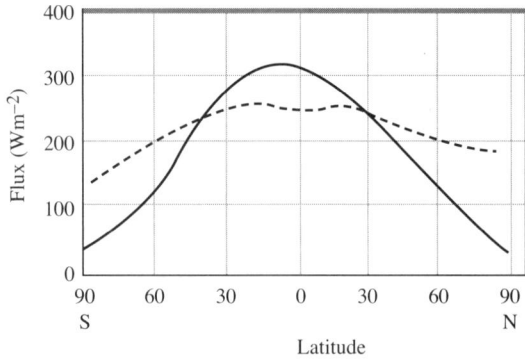

FIGURE 2.12   Heating and cooling rates of the Earth-plus-atmosphere system for the equinoxes near March 21 and September 21. Solid curve: heating rate from absorption of sunlight. Dashed curve: cooling rate from loss of terrestrial radiation to space.

(cos θ) S_E, where θ is the angle between the local vertical and the direction of the incoming beam. Consequently, the energy absorbed per unit horizontal area of the Earth's surface decreases from equator to pole in a systematic way, or a fixed amount of solar energy per second spreads over a larger horizontal area as latitude increases toward the poles. Latitudinal variations in cloudiness can lead to structure, especially in the solar heating curve. In addition, the polar ice caps are efficient reflectors of sunlight, but the primary latitudinal variation in Figure 2.12 arises from the spherical geometry. In radiative equilibrium, the areas under the solar heating and terrestrial cooling curves are equal.

If a local radiative equilibrium applied, then a plot of longwave energy lost to space would coincide with the curve for solar heating at each latitude, but satellite measurements reveal a different situation. The line for cooling by terrestrial radiation displays less curvature with latitude than that for solar heating. In a globally averaged sense, the Earth-plus-atmosphere system is in radiative equilibrium, but a balance of heating and cooling does not prevail at each latitude. The area of the globe that experiences a net heating is counterbalanced by an equal area of net cooling. Within approximately 35° latitude of the equator, there is a net heating, while there is a net cooling at higher latitudes in both hemispheres.

Based solely on Figure 2.12, the low-latitude regions should be heating up over time, and the higher latitudes should be cooling down, but this scenario is incomplete. The missing process is transport of heat by winds and ocean currents. Motions of the atmosphere and ocean move heat, in the form of warm air or warm water, from low latitudes toward high latitudes, while cooler air or water makes the return trip. These circulations act to cool the tropical regions and warm the higher latitudes. The results are tropical regions that are cooler than they would otherwise be and high latitudes that are warmer than if radiative equilibrium existed on a latitude-by-latitude basis. Chapter 4 considers the winds that develop in response to this latitudinal imbalance between heating and cooling. These motions can become very complicated, but their overall effect is to transport heat from equator to pole.

People who live at middle latitudes are keenly aware of the changing seasons over the course of a year. Figures 2.12 and 2.13 combined with some geometry clarify the origin of the seasonal cycle. First, the plane defined by the Earth's equator is tilted with respect to the plane formed by the Earth's orbit around the sun. The angle between the two planes is approximately 23.5°. Second, the Earth's rotation axis always points in the same direction as the planet orbits the sun. Near December 21 and June 21 of each year, the Earth–sun geometry is as shown in Figure 2.14. Near December 21 the sun, when at its highest point in the sky, is directly overhead at latitude 23.4–23.5° south, while near June 21 this situation exists at latitude 23.4–23.5° north. The former condition defines summer solstice in the Southern Hemisphere,

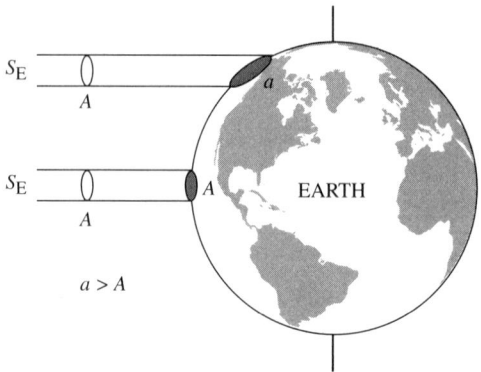

FIGURE 2.13   Incoming solar energy strikes a spherical Earth. A fixed amount of solar energy per unit time is spread over a different horizontal area depending on latitude. The energy received per unit horizontal area declines from the equator toward the poles.

and the latter is the start of summer in the Northern Hemisphere. The intermediate situations, in which the sun is overhead at noon at the equator, occur near March 21 and September 21. In December, the latitudinal distribution of solar heating is similar to that shown in Figure 2.12 except that the maximum shifts to latitude 23.4–23.5° south, and in June the maximum heating is at 23.4–23.5° north.

As the solar heating rate varies systematically over the course of a year at any fixed latitude, so does the surface temperature. One can view temperature as continually changing in an attempt to establish a longwave cooling at each latitude that balances the shortwave heating, although transport of heat by motions complicates this picture. At tropical latitudes, the solar heating rate remains large all year as the latitude of the maximum drifts between

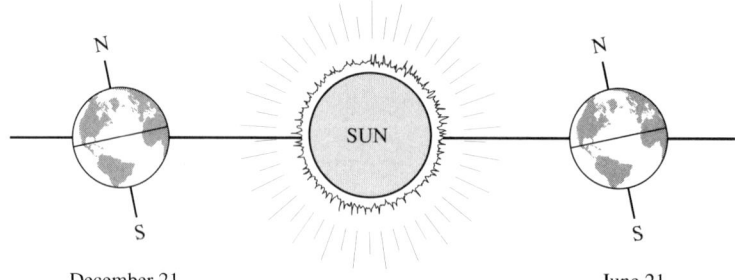

December 21   June 21

FIGURE 2.14   Orientation of the Earth and sun for dates near December 21 and June 21 of each year. Near December 21 the sun is directly overhead at noon at latitude 23.4–23.5° south. Near June 21 this occurs at latitude 23.4–23.5° north.

about 23.5° south and 23.5° north. The near-constancy of solar heating in this region leads to a surface temperature that varies little over the course of a year. As one proceeds toward either pole, the amplitude of the annual cycle in solar heating increases, and the temperature behaves in a similar fashion. The result is an annual cycle in temperature, called the seasonal cycle, in which the variation becomes greater with increasing latitude.

2.10   Exercises

1. The planet Venus is totally covered by very thick clouds that reflect incoming sunlight back into space. As a consequence, the albedo of Venus is about $A = 0.8$. The distance of Venus from the sun is $L = 1.08 \times 10^{11}$m.

   a. What is the effective radiating temperature of Venus?

   b. Venus is closer to the sun than is the Earth, so the incident solar energy is greater on Venus. Yet, the effective radiating temperature for Venus is colder than that for Earth. What is the origin of this difference?

   c. The surface temperature of Venus is in the vicinity of 750 K, much higher than the planet's effective radiating temperature. The atmosphere of Venus is composed almost entirely of carbon dioxide plus thick clouds containing sulfuric acid. What must the longwave extinction thickness of the atmosphere be to account for the high surface temperature?

2. One of the complications in the theory of climate involves a possible change in cloudiness in response to a change in the temperature of the Earth's surface. Clouds cover approximately 50% of the Earth at any given time. The "fractional cloud cover" is $f_c = 0.5$. The albedo of the present-day planet is $A = 0.3$. If there were no clouds at all ($f_c = 0.0$), the albedo would be close to $A = 0.1$. As a simple approximation, one can assume that fractional cloud cover and global albedo are related by the equation $A = 0.10 + 0.40 f_c$. As a result of the current greenhouse effect, the globally averaged surface temperature of the Earth is 33 K warmer than the effective radiating temperature.

   a. Now assume that the Earth experiences a greenhouse warming, and as a result the surface temperature becomes 36 K warmer than the effective radiating temperature. In addition, fractional cloud cover increases from its current value of $f_c = 0.5$ to $f_c = 0.6$. What is the new average surface temperature of the Earth?

   b. As in (a), assume that the surface temperature is 36 K warmer than the effective radiating temperature but that the fractional cloud cover decreases from $f_c = 0.5$ to $f_c = 0.4$. What is the new average surface temperature of the Earth?

3. Stars go through a life cycle during which their energy output changes slowly with age. One of the predictions from astrophysics is that the energy output of the young sun, 3 billion years ago, was 30% less than it is today. For the following part (a), assume that 3 billion years ago the albedo of the Earth-plus-atmosphere system was $A = 0.3$. Assume in both (a) and (b) that the magnitude of the greenhouse effect 3 billion years ago was the same as on the modern Earth, where the average surface temperature is 33 K warmer than the effective radiating temperature.

   a. Simple life forms existed on Earth 3 billion years ago, and these life forms required liquid water. Is this biological fact consistent with the expected average surface temperature as of 3 billion years ago? Support the answer with a quantitative argument.

   b. Suppose one requires the average surface temperature of the Earth to equal or exceed the freezing point of water, 273 K. Is it possible to resolve the difficulty revealed in (a) by claiming that the young Earth's albedo differed from the modern-day value of $A = 0.3$? Carry out a quantitative evaluation to support a yes or a no answer.

   c. Suggest one way to resolve the difficulty other than changes in the Earth's albedo. In this answer, relax the assumptions made in (a) and (b).

4. The atmosphere of a planet contains "one extinction thickness" of a greenhouse gas that absorbs and emits at all wavelengths in the longwave part of the spectrum ($\tau = 1$). Shortwave solar radiation is incident on the planet, and the "solar constant" is $S = 1300$ W m^{-2}. When solar energy interacts with the planet-plus-atmosphere system, 50% of the energy is absorbed at the planet's surface, 20% is absorbed in the planet's atmosphere, and 30% is reflected back to space. What is the globally averaged temperature of the planet's surface (T_G)?

5. The atmosphere of Planet X contains one extinction thickness ($\tau = 1$) of a greenhouse gas and clouds that absorb terrestrial radiation at all wavelengths over the longwave part of the spectrum. The albedo of the planet-plus-atmosphere system is $A = 0.4$, and the incoming shortwave stellar energy flux is $S = 1600$ W m^{-2}. The atmosphere is transparent to shortwave radiation, so all of the absorption of stellar energy takes place at the ground. The temperature of the ground is T_G and the temperature of the atmosphere (viewed as a single layer) is T_A.

   When stellar radiation is absorbed at the ground, some of the energy goes into evaporating liquid water from the surface of Planet X. Vertical motions transport this water vapor into the atmosphere, where it condenses back to liquid water to form clouds. In the process of condensation, energy (called "latent heat") is released and acts to warm the atmosphere. This sequence of events (heating of the ground, evaporation of water, transport of water vapor into the atmosphere, and latent heat release in condensation) effectively removes energy from the ground and adds it to the atmosphere. One can describe these processes by a flux of energy, ϕ in W m^{-2}, from the planet's surface to the atmosphere. This is called the "flux of latent heat."

Suppose a fraction β (where $0 < \beta < 1$) of the total radiant energy (shortwave plus longwave) that is absorbed at the ground goes into evaporating water, and this energy is then transported into the atmosphere to be released as latent heat. Given this, the energy per second in the form of latent heat transferred from ground to atmosphere is

$$4\pi R^2\phi = \beta[E(\text{short}) + E(\text{long})]$$

where E(short) and E(long), respectively, are the shortwave and longwave energies per second (in watts) absorbed at the ground, summed over the entire planet. The radius of the planet is R.

The total system of Planet X plus its atmosphere is in radiative equilibrium. The only way energy enters or leaves the system is by radiation. However, the ground alone, or the atmosphere alone, is in a state of "radiative-convective equilibrium," in which energy can move from one part of the system to another by radiation and by motions that are accompanied by latent heat release.

a. Let $\beta = 0.30$, so that 30% of the radiant energy absorbed by the ground goes into evaporating water and is eventually released as heat to the atmosphere. What are (1) the effective radiating temperature T_E of the Planet X-plus-atmosphere system, (2) the temperature T_A of the atmosphere, and (3) the temperature T_G of the ground? Derive the appropriate algebraic expressions for these quantities and then compute numerical values.

b. Compare the result from (a) for the atmosphere's temperature (T_A) to the analogous result for the case in which there is no flux of latent heat from the ground to the atmosphere ($\phi = 0$). If the results for the two cases differ, give a clear physical explanation for why they are different. If they are the same, give a clear physical explanation as to why they are the same.

c. Compare the result from (a) for the surface temperature (T_G) to the analogous result for the case in which there is no flux of latent heat from the ground to the atmosphere ($\phi = 0$). If the results for the two cases differ, give a clear physical explanation for why they are different. If they are the same, give a clear physical explanation for why they are the same.

2.11   References

Bohren, C. F., and D. R. Huffman. 1983. *Absorption and Scattering of Light by Small Particles.* New York: John Wiley and Sons.

Christiansen, E. H., and W. K. Hamblin. 1995. *Exploring the Planets.* Englewood Cliffs, N.J.: Prentice–Hall.

Feynman, R. P., R. B. Leighton, and M. Sands. 1964. *The Feynman Lectures on Physics.* Vol. 2. Reading, Mass.: Addison-Wesley.

Gibson, E. G. 1973. *The Quiet Sun.* Washington, D.C.: U. S. Government Printing Office.

Goody, R. M., and J. C. G. Walker. 1972. *Atmospheres.* Englewood Cliffs, N.J.: Prentice–Hall.

Gough, D. 1977. Theoretical predictions of variations of solar output. In *The Solar Output and Its Variation,* O. R. White, ed. 449–74. Boulder: Colorado Associated University Press.

Hanel, R., C. Prabhakara, B. J. Conrath, V. G. Kunde, I. Revah, V. V. Salomonson, and G. Woodford. 1972. The Nimbus 4 Infrared Spectroscopy Experiment: IRIS-D, I. Calibrated thermal emission spectra. *J. Geophys. Res.* 77:2629–41.

Hecht, E. 1996. *Physics: Calculus.* Pacific Grove, Calif.: Brooks Cole Publishing Co.

Herzberg, G. 1945. *Atomic Spectra and Atomic Structure.* New York: Dover Publications.

Houghton, J. T., L. G. Meira Filho, B. A. Callandar, N. Harris, A. Kattenberg, and K. Maskell, eds. 1996. *Climate Change 1995—The Science of Climate Change.* Cambridge, U.K.: Cambridge University Press.

Lean, J., and D. Rind. 1998. Climate forcing by changing solar radiation. *J. Climate.* 11:3069–94.

Liou, K. N. 2002. *An Introduction to Atmospheric Radiation.* New York: Academic Press.

Neckel, H., and D. Labs. 1984. The solar spectrum between 3300 and 12500 Angstroms. *Solar Physics.* 90:205–58.

Planck, M. 1959. *The Theory of Heat Radiation.* Translated by M. Masius. New York: Dover Publications.

U.S. Standard Atmosphere, 1976. Washington, D.C.: U.S. Government Printing Office, 1976.

Van der Waerden, B. L. ed. 1967. *Sources of Quantum Mechanics.* New York: Dover Publications.

Vonder Haar, T. H., and V. E. Suomi. 1971. Measurement of the earth's radiation budget from satellites during a five-year period. *J. Atmos. Sci.* 28:305–14.

Wehrli, C. 1985. *Extraterrestrial Solar Spectrum.* Publication 615. Davos, Switzerland: Physico-Meteorological Observatory, World Radiation Center.

Willson, R. C. 1997. Total solar irradiance trend during solar cycles 21 and 22. *Science.* 277:1963–65.

Chapter 3

Atmospheric Water

INTRODUCTION

Water may exist as a gas, liquid, or solid under the conditions found on Earth, and transitions between these phases have a profound influence on the state of the atmosphere. Attractive forces between molecules lead to a temperature-dependent upper limit on the amount of water that can exist in the vapor phase. A change in phase from liquid water or ice to vapor requires the addition of a specific amount of energy per unit mass to overcome intermolecular attractions, while the inverse transition releases the same quantity of energy, referred to as latent heat.

When a volume of air rises and cools, its relative humidity can increase to the point where condensation or freezing occur. The consequences are the formation of cloud droplets and the release of latent heat that warms the atmosphere. The temperature structure of the global troposphere reflects the net effect of rising and sinking air parcels with the accompanying release of latent heat. Once cloud droplets form, larger rain drops may grow and produce precipitation. All of these processes work together to determine the global cycling of water in which air motions, the flow of rivers, and evaporation, condensation, and precipitation move water between the land, oceans, and atmosphere.

3.1 Water on Earth and in the Atmosphere

The water molecule has a prominent place in the discussion of radiation in Chapter 2. Condensed water in the form of clouds is a major contributor to the Earth's albedo, and water vapor is the single most important greenhouse gas in the atmosphere. Water influences atmospheric temperatures in another way unrelated to solar and terrestrial radiation. When water vapor changes phase to liquid or ice to form clouds, heat is released to the troposphere, and this heating leads to much of the lower atmosphere being warmer than would otherwise be the case. Water occupies a unique role in the climate system, and this chapter focuses exclusively on the properties of this important molecule.

Water in the Earth-plus-atmosphere system is distributed among the three major reservoirs listed in Table 3.1. The oceans are by far the largest reservoir, with a volume of approximately 1.39×10^9 km^3, containing about 96.5% of the total mass of water (Gleick 1996). The land reservoir includes lakes, rivers, underground water, and ice on top of the land, where Antarctica is the most prominent example. The land-based reservoir contains approximately 3.5% of the Earth's water. Estimates of the partitioning of water among the reservoirs vary somewhat because of uncertainties in the quantity of groundwater. For example, Botkin and Keller (1998) list 97.2% as the percentage of Earth's total water in the oceans, with only 2.8% on land. In any case, the atmosphere contains only about 1/1000th of 1% of the water on Earth. This number includes water vapor as well as the liquid water and ice in clouds. Irrespective of this small percentage, atmospheric water is a major contributor to maintaining the Earth's climate in the range that allows life to exist.

The unusual property of water is that it can exist in any of three different phases under conditions found in the Earth-plus-atmosphere system. These are (1) gas or vapor, (2) liquid, and (3) solid or ice. Many substances exhibit three phases; for example, carbon dioxide freezes into a solid, called "dry ice," if cooled sufficiently. Even molecular nitrogen and oxygen can become liquids, but they have to be cooled to extremely low temperatures first. The distinguishing trait of water is that it undergoes transitions between the vapor, liquid, and solid phases at temperatures found on the Earth's surface and in the atmosphere.

How can one describe the water vapor content of a volume of air? The most straightforward way is to state the number density of water vapor; this is simply the number of water molecules per unit volume. The number density of a specific molecule is usually denoted by the chemical symbol in square brackets, in this case [H$_2$O]. Another way to measure the water vapor amount is by the pressure exerted by water molecules, denoted P_{H_2O}. Chapter 1 described the ideal gas law and applied it to the atmosphere as a whole. This is $P = nkT$, where k is Boltzmann's constant, T is absolute temperature, and n is the total atmospheric number density. If the gas is a mixture of different molecules such as N$_2$, O$_2$, and H$_2$O, then each individual component obeys its own gas law. Applied to water vapor this is

$$P_{H_2O} = [H_2O]\, kT \qquad [3.1.1]$$

TABLE 3.1 Water Storage on Earth

Reservoir	Percent of Total Water
Oceans	~96.5%
Land	~3.5%
Atmosphere	~0.001%

where it is possible to show that the same temperature applies to each component of the gas. The quantity P_{H_2O} is the "water vapor pressure," defined as the force exerted on a unit area by the water vapor molecules that strike the surface. Number density and vapor pressure are both reasonable ways to describe the water vapor abundance.

Before proceeding, it is worthwhile to note that the ideal gas law contains an important assumption. This simple result ignores any attractive or repulsive forces between molecules. The ideal gas law implicitly assumes that one molecule interacts with another only when they collide. In the case of water vapor, this assumption is not valid under all environmental conditions. Attractive forces exist between water molecules, and these forces become important when the temperature is sufficiently low and the number density is sufficiently high. These are the conditions under which a phase change can occur. Eq. 3.1.1 breaks down if applied to conditions where water vapor is close to condensing to a liquid. In this case, an "equation of state" still exists, but it is more complicated than Eq. 3.1.1. Fermi (1956) discusses modifications to the equation of state when the gas is in the regime where a change in phase can take place.

The next way to measure the water vapor content of air is the "mixing ratio," denoted by χ. This is the ratio of the number of H_2O molecules in a unit volume of air to the total number of all molecules irrespective of their identity. If the ideal gas law is valid, this is the same as the ratio of the water vapor pressure to the total atmospheric pressure:

$$\chi = [H_2O]/n = P_{H_2O}/P \qquad [3.1.2]$$

If conditions are such that the water vapor is close to condensing to a liquid, these two expressions are not numerically equivalent since the ideal gas law no longer applies.

Weather reports say nothing about water vapor number density, pressure, or mixing ratio. Instead, the water vapor amount is reported as the "relative humidity." It is expressed as a percentage, but a percentage of what? To understand the meaning of relative humidity, it is necessary to consider the ability of water to change phase. Imagine a volume of air that contains no water vapor whatsoever. Initially the air is totally dry, but molecules of water vapor are gradually added to the volume. The process of adding water vapor continues until a certain critical point is reached. Once the water vapor number density or vapor pressure grows to a specific value, the water vapor starts to undergo a change in phase. It begins to condense to a liquid, or if the temperature is sufficiently low, it freezes to ice. When this specific water vapor pressure is reached, if 1 million more molecules in the vapor state are added to the volume, then 1 million molecules of H_2O will condense out as liquid.

This thought experiment leads to the following conclusion: Only a certain maximum amount of water in the vapor phase can exist in a volume.

Once this maximum amount is reached, any additional water vapor will condense out as liquid. This condition, where a volume contains the maximum quantity of water vapor, is called "saturation," and the critical water vapor abundance at saturation is called the "saturation vapor pressure," denoted by P_{SAT}. Alternately, the mixing ratio of H_2O at saturation is the "saturation mixing ratio," χ_{SAT}. When air is saturated, any further water vapor added to the volume will condense out to liquid (at least in principle), or if the temperature is below 273 K, it freezes to ice. In fact, observations reveal a more complicated situation than indicated here. Water vapor in excess of the saturation amount can exist in a volume, whereas liquid water can continue to exist below the expected freezing point. Section 3.9 considers these issues further. For the present discussion, it is sufficient to adopt the view that saturation represents an inviolate upper limit.

There is one layer of complication to add to the above discussion; specifically, saturation vapor pressure depends on temperature. As temperature rises, a greater quantity of water can exist in the vapor phase; that is, P_{SAT} and χ_{SAT} increase rapidly as temperature increases. This dependence has an observable effect on the weather. In the hot summer, more water vapor can reside in a volume of air than during the cold winter. At low temperatures, the saturation vapor pressure is small, and this fact influences the magnitude of wintertime snowfalls. A volume that is colder than, say, 255 K or about 0°F cannot contain enough water vapor to produce a major snow. Consistent with this, the biggest snowfalls tend to occur when the temperature is only a little below freezing.

Table 3.2 illustrates the dependence of saturation vapor pressure on temperature. The "millibar" (mb) is a convenient unit for measuring atmospheric pressure, where the total pressure of the Earth's atmosphere at sea level is approximately 1013 mb *(U.S. Standard Atmosphere, 1976)*. The relationship between the millibar and the standard metric unit is 1 mb = 100 Pa = 100 N m^{-2}. Saturation vapor pressure P_{SAT} specifies the maximum contribution that water vapor can make to the total. At 273 K, or 32°F, water vapor can contribute a maximum of about 0.6% of the total pressure of

TABLE 3.2 Saturation Vapor Pressure of Water at Several Temperatures

T (K)	T (F)	P_{SAT} (mb)
273	32	6.11
283	50	12.27
293	68	23.37
303	86	42.38
313	104	73.78
323	122	123.40

the atmosphere. On an extremely hot day, the water vapor pressure might be 10 times greater, up to 6% to 7% of the total atmospheric pressure. This is why the tropical atmosphere contains more water vapor than that at middle latitudes, at least when averaged over a year.

It is important to recognize that on any given day at any location, the actual abundance of water vapor might be much less than the saturation amount, or $P_{H_2O} \leq P_{SAT}$, where P_{H_2O} is the actual water vapor pressure. Saturation defines the upper limit on the vapor abundance, and the actual water vapor abundance at a location depends on the history of the air mass that happens to be there, including the influence of regional sources and sinks. For example, in a hot desert, like Death Valley, the saturation vapor pressure is large because the temperature is high. Although a large amount of water vapor can exist under these conditions, the actual water vapor pressure is usually very small. This is because moist air masses seldom flow into the region, and there are no local sources of water vapor. On the other hand, near an ocean, where a large amount of water is always available to the air, the actual water vapor pressure is likely to be much closer to the saturation value.

Given the concept of saturation, it is possible to define relative humidity. Relative humidity, denoted by r, is the ratio of the actual water vapor pressure to the saturation vapor pressure, with a factor of 100 included to convert the quotient into a percentage:

$$r = 100 \, P_{H_2O}/P_{SAT}(T) \qquad [3.1.3]$$

where the notation in the denominator of Eq. 3.1.3 includes the temperature dependence of saturation vapor pressure. Relative humidity is the actual water vapor pressure expressed as a percentage of the maximum possible vapor pressure at the local temperature. When relative humidity reaches 100%, the volume is saturated, and in principle water vapor can begin to condense to liquid or, at sufficiently low temperatures, to freeze to ice.

The quantity commonly stated in a weather report is relative humidity at or very near the ground, but r usually varies with altitude because both P_{H_2O} and temperature change with distance in the vertical. This fact is important for the formation of clouds and precipitation. The relative humidity at the ground can be less than 100%, but a couple of kilometers higher the air can be saturated. As a consequence, clouds are usually located above the ground. When the relative humidity at the ground reaches 100%, a fog may develop.

The relative humidity early in the morning is usually higher than that in the afternoon, but this does not necessarily mean that the air contains less water vapor later in the day. Instead, temperature typically rises from early morning to afternoon, so the saturation vapor pressure in the denominator of Eq. 3.1.3 increases. This leads to a decline in relative humidity if P_{H_2O} in the numerator remains fixed. Another implication of this temperature dependence is that a relative humidity of 50% on a hot day corresponds to a larger absolute water vapor pressure than a relative humidity of 50% on a cold day. Relative humidity always describes how close the true water vapor amount is to the saturation value. Once r reaches 100%, water vapor can begin a change in phase, subject to the complications identified in Section 3.9.

3.2 The Phases of Water

What processes occur on the molecular level when a gas converts into a liquid or a solid? To understand how the state of saturation comes about, it is necessary to consider the three phases of water on a microscopic scale. The important concept here is that "intermolecular forces" exist between water molecules. These forces are electrical, and they act to pull one water molecule toward another. Differences among the three phases of water arise from differing strengths of the intermolecular forces. A water molecule as a whole is electrically neutral; the positive charges of the protons exactly balance the negative charges of the electrons. Mathematically, the electrical force is analogous to the gravitational force considered in Chapter 1. The force F between two electrical charges q_1 and q_2 is

$$F = q_1 q_2 / r^2 \qquad [3.2.1]$$

where r is the distance between these charges. If q_1 and q_2 have the same sign, the force is repulsive, and when the charges have different signs, the force is attractive.

How does Eq. 3.2.1 pertain to water molecules? A water molecule consists of one oxygen atom bound to two hydrogen atoms. Originally the free oxygen atom possessed eight electrons and each of the hydrogen atoms contained one electron. When the H_2O molecule forms, some of the electrons are shared among the three atoms, and this sharing constitutes a chemical bond. On average, the shared electrons tend to stay closer to the oxygen atom than to the hydrogen atoms, so although the molecule as a whole is electrically neutral, the electrical charge is not distributed uniformly in space. Oxygen is "electronegative," meaning that it preferentially pulls the electrons toward itself. Consequently, the region of the molecule near the oxygen atom tends to be a bit negative, whereas the hydrogen atoms tend to be a bit positive. Figure 3.1 illustrates the separation of charge in a single water molecule. The uneven distribution of electrical charge within the individual molecules gives rise to forces between different water molecules, and Eq. 3.2.1 describes these. The hydrogen atom in one H_2O molecule is attracted to the oxygen atom of another, and this electrical attraction is strong enough that it can constitute a form of bonding between the molecules. Under cer-

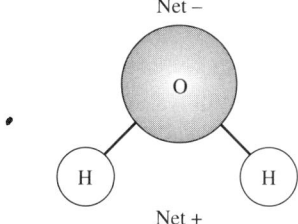

FIGURE 3.1　Regions with net positive and negative charges in the water molecule. The pull of the oxygen nucleus leads to a net negative charge on this end of the molecule, leaving the hydrogen atoms with a net positive charge.

tain conditions of temperature and number density, the forces cause large numbers of water molecules to clump together, resulting in a liquid or a solid.

The three phases of water are characterized by differing strengths of attraction between molecules. In the gas phase, individual molecules move about at random. The molecules are far enough apart, on average, that the attractive forces are negligible. In an "ideal gas," the molecules do not attract each other at all so, strictly, water vapor is not ideal. However, if the number density of the gas is sufficiently low and the temperature is sufficiently high, it is possible to neglect the attractive forces. Eq. 3.2.1 illustrates why. The attractive forces vary as the inverse square of the distance between molecules. At low number densities, the molecules are relatively widely spaced, so the forces are weak. Similarly, if temperature is high, molecules in the gas phase are moving rapidly. In this case, even when the molecules collide, they have sufficient kinetic energy that the intermolecular forces have no significant influence on their motion. Finally, because molecules move about individually, a gas will expand to fill any volume available to it. The following list summarizes these properties, which apply to any gas, not just to water vapor.

Properties of the Gas Phase

1. Negligible intermolecular forces (approximates an ideal gas).
2. Exists at low number density and high temperature.
3. Molecules move at random and interact only upon collision.
4. Expands to fill the available volume.

The liquid phase is the most difficult to understand in a rigorous way. The molecules in a liquid are packed closer together than those in a gas, and therefore the density of a liquid is greater than that of a vapor. Obviously, a glass of liquid water has a much greater mass than the same glass filled with water vapor. When the molecules are closely spaced, intermolecular forces come into play. In the liquid phase, attractive forces hold clumps of molecules together, but not so strongly that a fixed shape results. The following list refers to these as "moderate" attractive forces. In a liquid, these molecular assemblages tend to move about as separate entities, slipping over the surrounding molecular groups. Because these units are mobile, the shape of a liquid changes freely to fit any container.

Properties of the Liquid Phase

1. Moderate attractive forces—molecules attract each other, but forces do not lead to a fixed shape.
2. Molecules clump into groups that move independently of each other.
3. Shape changes freely to fit the container.
4. Must add energy (heat) to convert to the vapor phase.

Finally, it is necessary to add energy to a liquid to convert it into a gas. If sufficient heat is added to a container of liquid water, all of the liquid turns to vapor. One can view a liquid as being in a lower "potential energy" state than a gas, because energy is required to pull the molecules of a liquid away from each other to create the vapor. On a scale of potential energy, the vapor phase lies higher than the liquid phase. The term "potential energy" as used here might appear vague; all it means is that energy must be added to the liquid to convert it into a gas. This extra energy overcomes the attractive forces between water molecules in the liquid state.

In the solid phase, the attractive intermolecular forces are even stronger than in a liquid. The molecules in a solid are held in place, although they can vibrate back and forth around these positions. The result is that a solid, like an ice cube, maintains a fixed shape. Energy must be added to a solid to convert it into a liquid. When heat is added to an ice cube by taking it out of a freezer, it melts. Hence, solids lie even lower on the scale of potential energy than do liquids. The next list summarizes properties of the solid phase.

Properties of the Solid Phase

1. Strong attractive forces between molecules.
2. Molecules are held in place and vibrate around fixed positions.
3. Holds a fixed shape.
4. Must add energy (heat) to convert to a liquid.

The strength of the intermolecular forces is the essential difference between the three phases of water. All of the other points in the lists follow from this fact. Given information on how the phases differ, it is possible to return to the concept of saturation and describe the mechanisms by which this state comes about.

3.3   The Physical Basis of Saturation

Consider a closed container that is half filled with liquid water at a temperature T. Initially, the top half of the container is a vacuum; that is, there is no water vapor at all in the space above the liquid. Suppose that the container sits for several hours. Although a vacuum existed above the liquid initially, over time water vapor collects in the top of the container, leaving less liquid in the bottom. Figure 3.2 illustrates the situation. If ice existed in the container initially, the outcome would be qualitatively the same.

Molecules in the liquid phase possess internal energy. Although these molecules exist in collective units, individual molecules still have a limited degree of random thermal motion as the larger assemblages move about. Based on the previous discussion, energy must be added to molecules in the liquid to allow them to escape into the vapor phase. The experiment depicted in Figure 3.2 shows that some molecules originally in the liquid gained sufficient energy to jump into the vapor phase. A molecule has to gain energy to break away from the attractive forces in the liquid phase, but where does this energy come from? A fraction of the molecules in a liquid or a solid have relatively large amounts of internal energy that they gain via random

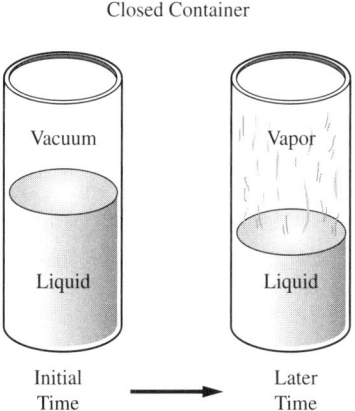

Closed Container

Vacuum   Vapor

Liquid   Liquid

Initial   Later
Time   Time

FIGURE 3.2   Evaporation of a liquid. Initially a vacuum exists above the liquid. Over time, a gas appears in this volume, as a fraction of the molecules overcome attractive forces in the liquid and make a transition to the vapor phase.

interactions with neighboring molecules. This energy exists in the form of vibrating or bending motions. The molecules in a liquid or solid have a "statistical distribution" of internal energies. An analogy to a statistical distribution involves the speeds of cars on an expressway. All cars do not travel at the same speed. A large portion of the cars along a highway move within ±5 miles per hour of the speed limit, but there is always a fraction moving much faster. Although there is an overall average speed that can be computed by considering all of the cars, some fraction of the total moves much faster than this average. Molecules behave in a similar way. They have a well-defined distribution of internal energies called a "Boltzmann distribution." The Boltzmann distribution has the following mathematical form:

[Fraction of Molecules with Internal Energy from E to $E + dE$]
$$= (kT)^{-1} \exp[-E/(kT)] \, dE \quad [3.3.1]$$

where E is the internal energy associated with molecular-scale motions. Because of the intermolecular forces in a liquid or solid, E must exceed a certain critical value before a molecule can break free and enter the vapor phase. At any given temperature, a fraction of the molecules in the liquid have enough energy to overcome the attractive forces and escape into the vacuum above the liquid in Figure 3.2. These energetic molecules that also reside near the liquid surface are the ones that collect as vapor in the top of the container.

When the container holds a liquid, the escape of energetic molecules is termed "evaporation" or "vaporization." When the container holds ice, the escape is called "sublimation." Sublimation may be less familiar than evaporation, but a simple experiment shows that the process occurs routinely. Simply leave an ice cube sitting in a freezer, and eventually it disappears. This may require several weeks, but it will happen given sufficient time. The temperature never goes above freezing, but the ice cube shrinks and vanishes anyway. At the fixed temperature in the freezer, a fraction of the ice phase molecules still acquire sufficient energy from their neighbors to escape into the vapor phase. If these molecules are not replaced by renewed freezing, the ice cube shrinks over time. Note that the air in the freezer does not become saturated provided its volume is sufficiently large and the door is opened and closed regularly to allow an exchange of air with the outside. In summary, evaporation and sublimation occur because some of the water molecules in the liquid or ice phase gain enough energy to overcome the attractive forces and escape into the vapor phase. This fraction is very dependent on temperature.

As the ice cube shrinks, it changes shape, and this is also a consequence of intermolecular forces. Suppose the cube starts out with sharp corners, as shown in Figure 3.3. A water molecule located at the upper center of the cube is surrounded by many neighbors, so it feels a strong attractive force.

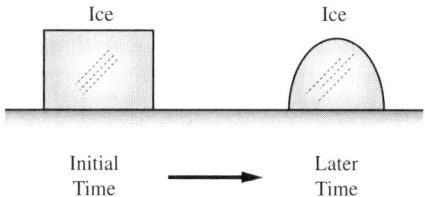

FIGURE 3.3   Sublimation of an ice cube. Molecules near the corners of the cube feel a weaker attractive force than do those in the upper center. Escape of molecules into the vapor phase leads to a rounding-off of the corners.

On the other hand, a molecule on the corner has fewer neighbors, so it feels less attractive force. Thus, it is easier for a molecule in the corner to escape into the gas phase because fewer molecules are holding it back. The result is that the sharp edges of the ice cube tend to become rounded as time passes. Eventually, it looks like the cube on the right of Figure 3.3.

The example discussed in connection with Figure 3.2 uses a closed container so that the vapor remains in contact with the liquid. As vapor accumulates in the upper portion of the container, some of the gaseous molecules collide with the liquid surface, stick to it, and enter the liquid phase. At this stage, molecules are leaving the liquid by evaporation, while other molecules are returning to the liquid. The more molecules that exist in the vapor phase, the more molecules there are to collide with the surface and return to the liquid. Eventually an equilibrium state develops in which the number of molecules that leaves the liquid per unit time equals the number that returns to the liquid. At this point, the water vapor pressure is as large as it can get, and this is the condition of saturation. In equilibrium, the water vapor pressure in the top of the container equals the saturation vapor pressure, or $P_{H_2O} = P_{SAT}(T)$. An alternate statement is that the relative humidity in the upper part of the container is 100%. This example also shows why saturation vapor pressure is a function of temperature. The warmer the liquid becomes, the greater is the average internal energy of its molecules. Via the Boltzmann distribution in Eq. 3.3.1, a larger fraction of the molecules have the energy needed to overcome the attractive forces in the liquid and escape into the vapor phase. So as the liquid becomes warmer, more molecules are able to collect as vapor in the upper part of the container. This is an alternate way of saying that saturation vapor pressure increases with temperature.

Imagine that the experiment in Figure 3.2 is done twice. In one case, the closed vessel contains liquid water and in the other case it contains ice, so the temperature is below 273 K. Also, assume that the vessel containing liquid water is at the same temperature as the ice, but the water has not yet frozen. This is called "supercooled water" and it is an important component of some, but not all, clouds. As before, vapor collects in the upper portion of both containers, and eventually the water vapor pressure reaches an equilibrium value equal to the saturation vapor pressure. However, measurements show that the saturation vapor pressure over the ice surface is smaller than the saturation vapor pressure over the liquid water surface at the same temperature. Mathematically, $P_{SAT}(T)_{ice} < P_{SAT}(T)_{liquid}$. More vapor collects over liquid water than over ice at same temperature. This observation illustrates the differing intermolecular forces in liquid and ice. The attractive force between molecules in the ice phase is stronger than the attractive force in the liquid phase. Hence, at a fixed temperature, fewer molecules have the energy needed to escape from the ice than from the liquid. To use a macroscopic analogy, it is easier for a person to get out of a bathtub filled with liquid water than from a bathtub in which the person is encased in a block of ice. The technical statement of this result is this: The saturation vapor pressure over ice is smaller than the saturation vapor pressure over liquid water at the same temperature. Section 3.9 shows that the difference in saturation vapor pressure over ice as compared to that over liquid water is important for the formation of precipitation in the Earth's atmosphere.

3.4 Latent Heats

The preceding example focused on molecules escaping from liquid water or ice by merit of having energies greater than the critical value needed to overpower the attractive forces at work. These energies are much larger than average values for typical earthly temperatures. It is possible to take this reasoning one step further. Instead of waiting for only the most energetic molecules to evaporate, one can add energy to a volume of liquid water by heating it. The additional energy increases the number of very energetic molecules, with the result that more molecules evaporate. Imagine adding just enough energy to 1 g of liquid water to make 100% of it evaporate, no more energy and no less. This specific amount of energy, denoted by L_v, is called the "latent heat of vaporization." On the molecular level, L_v is the energy needed to overcome all of the attractive forces that exist between the molecules in one gram of liquid water. The simple act of boiling water illustrates the process. Tiny bubbles appear in the bottom of a pot of water when it is heated sufficiently. Eventually the bubbles grow and rise to the surface of the liquid. These are not air bubbles. Instead, they are volumes of water vapor that form by a phase change in the hottest part of the liquid. When many bubbles are forming, the water is boiling.

This experiment can operate in reverse by cooling 1 g of water vapor. The cooling process involves taking energy away, for example, by putting the vapor in contact with a cold surface. If the vapor condenses to liquid, some energy in the form of heat is released to the environment. This energy per gram of water is called the "latent heat of condensation," L_c, and numerically it is identical to the latent heat of vaporization, $L_v = L_c$. In vaporization, energy is added to the liquid, and in condensation the same amount of energy is released to the surrounding air. The latter direction is important in the atmosphere, where the condensed liquid water forms a cloud, and the released latent heat of condensation goes into warming the air.

The value of the latent heat is approximately $L_v = L_c = 2.50 \times 10^6$ J per kg H_2O. To be more exact, the latent heat varies with temperature, where an approximate dependence is $L_v = L_c = (2.50 \times 10^6) - (2.37 \times 10^3)$ $(T - 273.15)$ J per kg H_2O, and temperature T is on the Kelvin scale (Hess 1959). The latent heat decreases as temperature increases, as one can infer on physical grounds. To understand this, consider evaporation. As the temperature of a liquid rises, an increasing fraction of the molecules in 1 g of water have the energy needed to escape into the vapor phase of their own accord, irrespective of any external heating. Hence, the extra energy required to get 100% of the liquid to evaporate must decrease as the temperature of the liquid increases.

This example considers transitions between the liquid and vapor phases. The other possibilities are changes in phase between ice and liquid and between ice and vapor. To melt an ice cube, heat must be added to it. One way to do this is to place the cube in contact with warm air. The energy required to melt 1 g of ice, converting it to liquid water, is called the "latent heat of melting," L_m, whose value to two significant figures is $L_m = 3.3 \times 10^5$ J per kg H_2O (Hess 1959). The addition of still more energy can convert the ice directly to vapor in the process of sublimation defined previously. From the standpoint of energetics, sublimation is a two-part process. First, energy in the amount required to convert from ice to liquid is expended, and then additional energy drives the transition directly to the vapor phase. Hence, the latent heat of sublimation, L_s, is the sum of the latent heats for melting and vaporization, $L_s = L_m + L_v = 2.83 \times 10^6$ J per kg H_2O, where these numerical values neglect temperature dependence. Of course, when these processes proceed in reverse, an equal amount of energy is liberated to the environment. These exchanges of energy all trace their origin to attractive electrical forces that exist between water molecules.

One can view ice as being in a lower potential energy state than liquid water, and the liquid phase lies still lower than water vapor. This hierarchy allows representation of the phases of water in terms of the energy level diagram in Figure 3.4. The values of the various latent heats define the energy separations between the phases. The latent heat of condensation is large relative to the energy required to change the temperature of a gram of liquid water by several degrees Kelvin. When water vapor condenses to liquid, the release of latent heat leads to a substantial warming of the surrounding air, and it thereby has a major influence on the temperature structure of the troposphere.

For future applications, it is necessary to put the analysis of latent heats on a quantitative basis. Consider a volume V of air that contains water

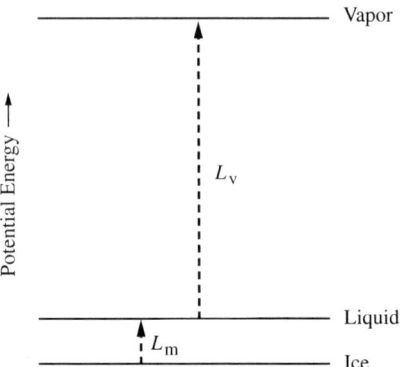

FIGURE 3.4 Energy level diagram illustrating the relationship between the three phases of water. The latent heats for vaporization and melting define the separations between energy states.

vapor mixed with the other major gases, as depicted in Figure 3.5. Initially, a mass of water vapor M_{VAP} resides in this volume, where the subscript VAP indicates water in the vapor phase only. Let the total mass of air in the volume be M_{AIR}. Most of this mass consists of N_2 and O_2, but M_{AIR} includes the water vapor as well. Now suppose that a mass of water vapor dM_{VAP} condenses to liquid. The latent heat of condensation specifies the amount of heat released to the air when this particular amount of vapor condenses. The energy released is dH, where

$$dH = -L_c \, dM_{VAP} \tag{3.4.1}$$

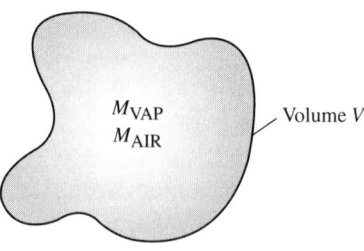

M_{VAP} = Mass of H_2O Vapor in V

M_{AIR} = Mass of Air in V

FIGURE 3.5 A volume of air in which water vapor is condensing to liquid and releasing latent heat. The latent heat released per unit mass of air is proportional to the change in water vapor mixing ratio.

The minus sign is necessary because $dM_{VAP} < 0$ since vapor is vanishing, but dH is positive when heat is released to the air. Eq. 3.4.1 serves as a mathematical definition of the latent heat of condensation.

Atmospheric applications seldom deal with well-defined volumes. In these cases it is convenient to think in terms of the heat released per unit mass of air. This is defined as $dq = dH/M_{AIR}$, typically expressed in ergs per gram of air. From Eq. 3.4.1,

$$dq = -L_c \, dM_{VAP}/M_{AIR} \qquad [3.4.2]$$

The ratio of masses can be expressed in a more useful form. The total mass of air is the mass per unit volume multiplied by the volume, $M = m \, n \, V$, where m is the mean molecular mass and n is the total number density of air in volume V. The mean molecular mass, defined in Chapter 1, includes the relative proportions of all molecules present, including water vapor. Similar reasoning applies to the evaluation of dM_{VAP}. The appropriate expression is

$$dM_{VAP} = m_{H_2O} \, d[H_2O]V \qquad [3.4.3]$$

where m_{H_2O} is the mass of water molecule, 2.99×10^{-23} g, and $d[H_2O] < 0$ is the change in water vapor number density as a result of condensation to liquid. With the preceding expressions, Eq. 3.4.2 becomes

$$dq = -L_c \, (m_{H_2O}/m)(d[H_2O]/n) \qquad [3.4.4]$$

where the ratio of molecular masses can be relabeled as $\varepsilon = m_{H_2O}/m$. The exact value of ε depends on the contribution of water vapor to m, but an approximate value is $\varepsilon = 0.622$, which applies to dry air. For practical purposes, the loss of a portion of the water vapor in condensation has an insignificant impact on m and n, so it is acceptable to assume that these quantities remain constant.

The water vapor mixing ratio is defined as $\chi = [H_2O]/n$, so the quantity $d[H_2O]/n$ is the change in water vapor mixing ratio due to condensation, $d\chi = d[H_2O]/n < 0$. Finally, the heat released per unit mass of air when water vapor condenses is

$$dq = -\varepsilon \, L_c \, d\chi \qquad [3.4.5]$$

where the minus sign indicates that a decrease in water vapor mixing ratio causes a release of heat to the surrounding air; $d\chi < 0$ implies $dq > 0$. Eq. 3.4.5 will prove useful in a theory to describe the temperature structure of the troposphere. This is the topic of Section 3.7.

3.5   Saturation Vapor Pressure: A Quantitative Expression

The concept of latent heat provides the basis for a mathematical expression that specifies saturation vapor pressure as a function of temperature. The derivation is beyond the scope of this text; details appear in Bohren and Albrecht (1998). The result, Eq. 3.5.1, provides useful insights into the mechanisms at work:

$$P_{SAT}(T) = P_{SAT}(T_0) \exp[-(m_{H_2O}L_v/k)(1/T - 1/T_0)] \qquad [3.5.1]$$

where the derivation assumes L_v to be independent of T. Eq. 3.5.1 relates $P_{SAT}(T)$ at any temperature to a known reference value $P_{SAT}(T_0)$ at temperature T_0, where T_0 must equal or exceed 273 K. All quantities in Eq. 3.5.1 have been defined previously. For $T_0 = 273$ K, the measured saturation vapor pressure is $P_{SAT}(T_0) = 6.11$ mb. Use of the known numerical values in Eq. 3.5.1 produces

$$P_{SAT}(T) = 6.11 \exp[-5417(1/T - 1/273)] \qquad [3.5.2]$$

where T is in Kelvin and $P_{SAT}(T)$ is in millibars (Bohren and Albrecht 1998). The exponential in Eq. 3.5.2 is a very sensitive function of temperature, a dependence that arises from the attractive forces between water molecules. For a physical interpretation, consider the portion of Eq. 3.5.1 that depends on temperature, $\exp[-m_{H_2O}L_v/(kT)]$. The quantity kT in the denominator measures the average thermal energy of a molecule in the liquid state, while $m_{H_2O}L_v$ in the numerator measures the energy needed for one H_2O molecule to break away from the attractive forces that hold it in the liquid phase. The argument in the exponential is the ratio of these two energies, and this has a clear meaning based on the Boltzmann distribution in Eq. 3.3.1. The exponential measures the fraction of molecules that have sufficient thermal energy to escape the attractive forces of the liquid and enter the gas phase. At typical atmospheric temperatures this fraction is very small, but it increases rapidly as temperature increases.

Eqs. 3.5.1 and 3.5.2 refer to the saturation vapor pressure over liquid water. If temperature is below 273 K, the saturation vapor pressure over ice results from using the latent heat of sublimation, L_s, in place of the value for vaporization, L_v. In this case, the value 5417 in Eq. 3.5.2 is replaced by 6143. This result, and especially the sensitive temperature dependence, is important in understanding the temperature structure of the troposphere and the formation of clouds.

3.6   Cloud Formation and Cloud Types

The concepts of saturation and changes in phase allow for a qualitative explanation of how clouds form. The definition of relative humidity,

$r = 100\, P_{H_2O}/P_{SAT}(T)$, implies that there are two ways to increase this quantity in a volume of air. One is to add more water vapor to the volume, thereby increasing the numerator, P_{H_2O}. The other way is to lower the temperature. Cooling the air causes the saturation vapor pressure, $P_{SAT}(T)$, to shrink and consequently the relative humidity increases, even when the absolute water vapor pressure remains fixed. If the temperature of a volume of air drops sufficiently, the relative humidity will reach 100%, and condensation can begin.

Rising motions are an effective way to lower the temperature of a volume of air. When a parcel of air rises in the atmosphere, its volume expands and its temperature drops in the process of expansional cooling introduced in Chapter 1. Conversely, when an air parcel sinks, its volume decreases and it becomes warmer. Section 3.7 puts these statements on a rigorous mathematical basis. Here it is sufficient to accept them as observational facts. A rising air parcel carries water vapor with it. If the temperature drop in the air parcel is sufficiently large, the relative humidity will reach 100% because of the shrinking saturation vapor pressure. At this point, the vapor can begin to condense to liquid or freeze. When this sequence of events occurs in the atmosphere, the result is a cloud. The classic text by Mason (1957) and the more recent work by Rogers and Yau (1989) review a broad range of topics related to cloud formation.

There is a subtle point in the preceding description. To see this, assume that the ideal gas law applies to the water vapor in a rising parcel, $P_{H_2O} = [H_2O]kT$. Strictly, this is valid only when the relative humidity is lower than 100%, but the argument made here applies nonetheless. Temperature decreases as the parcel rises, and as the parcel expands, the number density of water vapor, $[H_2O]$, declines because the same molecules are distributed through a larger volume. Both of these factors cause P_{H_2O} to shrink as the parcel rises, so the numerator in Eq. 3.1.3 decreases. The only way the relative humidity can increase is if the denominator, $P_{SAT}(T)$, decreases more than the numerator. This is where the sensitive temperature dependence of $P_{SAT}(T)$ is critical. If saturation vapor pressure were only weakly dependent on temperature, it would not be possible to create clouds by lifting and cooling a parcel of air. However, since the denominator in the definition of relative humidity decreases very rapidly as T decreases, the relative humidity can rise to 100%. Recall that the strong exponential temperature dependence of saturation vapor pressure arises from attractive forces between water molecules. These intermolecular forces have a direct influence on cloud formation.

A rising air parcel cools by 9.8 K for each 1-km increase in altitude provided water vapor is not yet condensing to liquid. This number describes the "dry adiabatic lapse rate," $\Gamma_d = 9.8$ K km^{-1}. The term "dry" means that water vapor is not condensing to liquid, and "adiabatic" indicates that no energy is being added to the parcel by external means. As used here, a positive value of the lapse rate corresponds to a decrease in temperature with increasing altitude. The lapse rate, $\Gamma_d = 9.8$ K km^{-1}, is valid only when the relative humidity in the parcel is less than 100%. When the parcel cools sufficiently, water vapor starts to condense, and latent heat is released to the air. At this point, for every kilogram of water vapor that turns to liquid, approximately 2.50×10^6 J energy are released. Two opposing effects operate here. Expansional cooling as the parcel rises acts to lower the temperature, but latent heat release tends to warm the air. The net result is that the air parcel still cools, but not as rapidly as before the vapor started to condense. The change in temperature per unit change in altitude is now described by a "moist lapse rate," denoted by Γ_m. There is no one universal value for the moist lapse rate since it depends on the amount of latent heat released as the parcel rises. If very little H$_2$O is condensing, the moist lapse rate will be almost the same as the dry adiabatic value. If, on the other hand, a large amount of water vapor is condensing, the moist lapse rate can be considerably smaller, where the minimum observed value is near 4 K km^{-1} (Wallace and Hobbs 1977).

Figure 3.6 depicts the temperature of a rising air parcel in an idealized manner to illustrate the processes at work. When the parcel begins its ascent near the ground, its relative humidity is $r < 100\%$, and its temperature cools by 9.8 K km^{-1}. As described, the cooling leads to an increase in relative humidity. Eventually r reaches 100%, at which point vapor begins to condense to liquid. The specific altitude at which relative humidity first equals 100% is called the "lifting condensation level" (LCL), and this corresponds to the bottom of a developing cloud. The temperature profile in the troposphere is determined by the dry adiabatic lapse rate, modified by release of latent heat as water vapor condenses to liquid. At sufficiently high altitudes, most of the

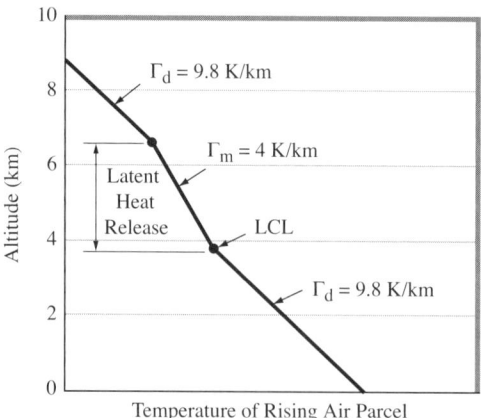

FIGURE 3.6 The relative humidity in an air parcel increases as the parcel begins to rise and cool. When the relative humidity reaches 100%, latent heat is released. The LCL is the altitude at which the relative humidity first reaches 100%.

water vapor that was initially in a parcel has condensed out to form a cloud. At this point, the temperature of a rising parcel once again changes at approximately the dry adiabatic value since ongoing condensation can release little latent heat. The observed average lapse rate for the entire troposphere is 6.5 K km^{-1} (*U.S. Standard Atmosphere, 1976*); if water vapor did not change phase, the value would be 9.8 K km^{-1}. The formation of clouds is a visible consequence of the release of latent heat, and on the basis of the preceding discussion, the existence of clouds is closely coupled to the temperature structure of the troposphere.

The lifting condensation level occurs at different altitudes for air parcels with different initial conditions. If a parcel starts out at the ground with a low relative humidity, it must ascend a long distance before it cools sufficiently for r to reach 100%. In this case, the LCL might be high in the troposphere. If the relative humidity in the parcel at the ground is high to begin with, the parcel needs to rise only a short distance before it cools enough for r to reach 100%. Here the LCL will be at a low altitude. Because of these differences associated with the initial humidities of the air parcels, latent heat is released over a wide range of altitude, and clouds can form over a wide vertical range in the troposphere.

Clouds display different physical traits depending on details of the vertical motions that create them. Low-pressure weather systems, discussed in Chapter 4, are areas where air ascends slowly over large horizontal areas, with characteristic dimensions as great as several hundred kilometers. The clouds that form in these regions are called "stratus." Stratus are characterized by large horizontal dimensions, roughly uniform coverage of the sky, and, occasionally, extended periods of precipitation.

Thermal convection, in which identifiable air parcels rise and water vapor condenses, produces clouds that are small in the horizontal dimension, with a characteristic scale on the order of a kilometer, but that can be large in vertical extent. In extreme cases, these "cumulus" clouds can reach and sometimes penetrate the tropopause. The strong updrafts that lead to vertically extended cumulus towers can produce thunderstorms, particularly after intense heating of the ground by sunlight.

The final cloud type, "cirrus," forms only at high altitudes, often near the tropopause. Since cirrus exist at very cold temperatures, the water consists entirely of ice crystals. In some cases cirrus are sufficiently thin that it is possible to see the diffuse blue light of the sky coming from above them. Cirrus typically measure on the order of a kilometer in the vertical and can cover tens of kilometers in the horizontal. Ice crystals blown from the tops of cumulus are the source of some cirrus clouds. In addition, high-flying aircraft can introduce both water vapor and small particles that act as sites for freezing into the atmosphere near the tropopause. There is observational evidence that cirrus coverage has been increasing over time near heavily traveled air corridors (Penner et al. 1999).

3.7 The Tropospheric Lapse Rate

This description of cloud formation is qualitative, and two important numerical values are stated without proof. Specifically, in the absence of latent heat release the tropospheric lapse rate would be 9.8K km^{-1}, but the observed globally averaged lapse rate is 6.5 K km^{-1}. The combination of several pieces of information yields a theory capable of producing both of these numbers. These pieces are the ideal gas law, hydrostatic balance, and the First Law of Thermodynamics.

The ideal gas law specifies a relationship between atmospheric pressure P, number density n, and absolute temperature T. This is $P = nkT$, given in Chapter 1 as Eq. 1.8.1. If the mean molecular mass of the atmosphere is m, the ideal gas law becomes $P = \rho RT$, where $\rho = nm$ is the mass density and $R = k/m$ is the "gas constant" for air. The "specific volume," $\alpha = 1/\rho$, is defined as the volume occupied by a unit mass of air. With these definitions, the ideal gas law becomes

$$P\alpha = RT \qquad [3.7.1]$$

Hydrostatic balance describes the altitude dependence of atmospheric pressure or number density for a given temperature profile. This is $dP = -mng\,dz = -\rho g\,dz$ based on Eq. 1.9.8. In terms of specific volume, the expression for hydrostatic balance is

$$dP = -(g/\alpha)\,dz \qquad [3.7.2]$$

If the altitude increment dz is positive, the change in pressure is negative.

The goal is to determine the lapse rate of temperature, dT/dz, but equations 3.7.1 and 3.7.2 involve three independent variables, P, α, and T. The First Law of Thermodynamics provides the additional information needed to complete the derivation. Planck (1945) gave an insightful account of the First Law and its verification by empirical means. The First Law states that the total internal energy of a system is conserved, where the system of interest here is a parcel of air rising or sinking in the atmosphere. One way to add internal energy to an air parcel is to heat it by putting it in contact with a warm ground. Another way is for some water vapor to condense to liquid in the parcel, thereby releasing latent heat. In response to being heated, the volume of an air parcel expands. When the volume increases, the parcel must expend energy to push the surrounding atmosphere away. As an analogy, someone who is surrounded by a crowd of people can expand his personal space by pushing other people back, but this requires an expenditure of energy. The First Law of Thermodynamics expresses these ideas in a quantitative manner. The mathematical statement is

$$du = dq - dw \qquad [3.7.3]$$

where dq is the energy per unit mass added to an air parcel. For the present application this will be in the form of latent heat release. The quantity dw is the work done per unit mass of air in expanding against its surroundings. Finally, du is the change in internal energy per unit mass of air. Thermodynamics considers Eq. 3.7.3 to be a definition of internal energy. If some energy dq is added to an air parcel and the parcel does work dw in expanding, the difference between these two quantities represents a change in internal energy of the air parcel. Each of the quantities, du, dq, and dw, is related to variables such as pressure, temperature, and the latent heat released.

The work done by a unit mass of air when its volume expands, dw, can be expressed in terms of pressure P and specific volume α. Consider a volume of air whose mass is 1 g, subjected to an external pressure. Traditional thermodynamic derivations envision a piston applying pressure to a gas, but the system could also be a layer of air squeezed by the pressure of the atmosphere above it. Since the mass under consideration is 1 g, volume and specific volume are identical. Hence, initially the volume of the air is α. Figure 3.7 depicts the situation. The volume of the gas now increases, where this expansion could be in response to the addition of energy such as latent heat release. In response to this heat, the gas does work by pushing upward against the piston, or against the pressure of the atmosphere above it. The volume expands from α to $\alpha + d\alpha$. This behavior was first observed experimentally, although today it can be predicted from a molecular description of the gas. The gas expands against the pressure exerted from above and pushes the cylinder back a distance dz, as shown in Figure 3.7. Work is defined in terms of a force acting over some distance: Work Done = Force × Distance (Hecht

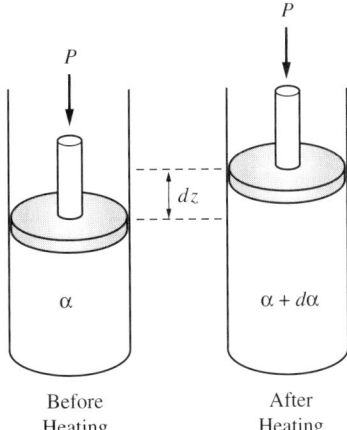

FIGURE 3.7   In response to addition of energy, a volume of air expands and warms while the pressure P remains constant. The work done is $dw = P\,d\alpha$, where the displacement dz of the piston leads to the change in volume $d\alpha$.

1996). In this example, force is the pressure multiplied by the cross-sectional area A of the volume of air, while the distance is the displacement of the piston: $dw = P A \, dz$. The product $A \, dz$ is the change in volume, $d\alpha$. The final result for the work done is

$$dw = P \, d\alpha \qquad [3.7.4]$$

Eq. 3.7.4 is a general result; it applies to pistons as well as to air parcels. To this point, the First Law of Thermodynamics can be expressed as

$$du = dq - P \, d\alpha \qquad [3.7.5]$$

The concept of internal energy is straightforward when one considers the molecular scale. One form of internal energy in a gas consists of the kinetic energy of molecules as they fly around at random, but internal energy also includes the vibrating and rotating motions of atoms that make up molecules. The temperature of a gas measures all of these forms of internal energy. A change in internal energy, du, is accompanied by a change in temperature, dT, and the two quantities are proportional. This proportionality was first established experimentally by James Joule in the middle of the nineteenth century (Planck 1945). Joule did a series of experiments to study the behavior of gases when they were heated. One of these experiments examined the relationship between the internal energy of dry air and its temperature. Joule added energy to a volume of air by heating it, but he held the volume constant. This can be done by keeping the air in a container whose volume is fixed. With reference to Eq. 3.7.5 this means that $d\alpha = 0$, so that $du = dq$. In this special case, the change in internal energy is equal to the energy added. Joule's experiments showed that the temperature change of the gas was proportional to the energy added: $dq = du \propto dT$. With a proportionality constant, the corresponding equality is

$$du = c_v \, dT \qquad [3.7.6]$$

where c_v is called the "specific heat at constant volume." These experiments held the volume of the gas constant, from which comes the name c_v. The measured value for air is $c_v = 0.717 \times 10^3$ J kg^{-1} K^{-1} (Wallace and Hobbs 1977). Joule did additional experiments to show that internal energy depends only on temperature, so the validity of Eq. 3.7.6 does not depend on the fact that volume was held constant. This result first came about empirically, but today it can be derived using a model of a gas where molecules move about, vibrate, rotate, and collide with each other at random.

Given Joule's result, the First Law becomes

$$c_v \, dT = dq - P \, d\alpha \qquad [3.7.7]$$

Eq. 3.7.7 can be applied to an air parcel in the atmosphere. If some energy dq is added to 1 g air at pressure P, its volume will change by $d\alpha$, and its temperature will change by dT. This straightforward reasoning will lead to the lapse rate of temperature in the troposphere.

The ideal gas law allows rewriting the last term on the right-hand side of Eq. 3.7.7. First, an identity from calculus is $P\,d\alpha = d(P\alpha) - \alpha\,dP$, but from the ideal gas law, $d(P\alpha) = R\,dT$. With these, Eq. 3.7.7 becomes $c_v\,dT = dq - R\,dT + \alpha\,dP$, or $(c_v + R)\,dT = dq + \alpha\,dP$. The quantity $c_v + R$ is called the "specific heat at constant pressure," $c_p = c_v + R$. The measured value is $c_p = 1.004 \times 10^3$ J kg^{-1} K^{-1} (Wallace and Hobbs 1977). If pressure is held constant, $dP = 0$, then $dq = c_p\,dT$, and hence the name c_p. The First Law of Thermodynamics is now

$$c_p\,dT = dq + \alpha\,dP \qquad [3.7.8]$$

Eq. 3.7.8 describes the relationship between changes in pressure and temperature when energy is added to a parcel of air. In the case of interest here, dq arises from release of latent heat in the condensation of water vapor.

Imagine an air parcel ascending through the atmosphere. Perhaps the air has encountered a mountain, and the parcel is pushed upward to get over the top. If a parcel is rising, the change in pressure it experiences, dP, is related to the change in altitude via hydrostatic balance, Eq. 3.7.2. This is where the link is made between the atmosphere and the First Law of Thermodynamics. The combination of Eqs. 3.7.2 and 3.7.8 yields

$$c_p\,dT = dq - g\,dz \qquad [3.7.9]$$

Eq. 3.7.9 says that if 1 g of air moves vertically a distance dz and is heated by an amount dq, it will experience a temperature change dT. This result applies whether the parcel is rising or sinking. Division of both sides of Eq. 3.7.9 by c_p and dz gives

$$dT/dz = -(g/c_p) + (1/c_p)\,dq/dz \qquad [3.7.10]$$

The lapse rate of temperature is defined as $\Gamma = -dT/dz$, where this sign convention gives a positive value of Γ when temperature decreases with altitude. The lapse rate is

$$\Gamma = (g/c_p) - (1/c_p)\,dq/dz \qquad [3.7.11]$$

This expression for the lapse rate arises by considering a single air parcel moving vertically through the atmosphere, where latent heat release provides the only addition of energy. A complete theory of the thermal structure of the troposphere would require inclusion of heating and cooling by radiative processes, discussed in Chapter 2, and horizontal transport of heat by winds, considered in Chapter 4. Still, to a first approximation, the temperature profile of the troposphere reflects the combined action of innumerable air parcels rising and sinking over the globe in the process of thermal convection.

Recall that dq is the amount of latent heat released in 1 g of air as the parcel moves a distance dz in the vertical, with water vapor condensing to liquid. If the air parcel has not reached saturation, then no water vapor will condense, so that $dq/dz = 0$ when $r < 100\%$. In this case, Eq. 3.7.11 becomes

$$\Gamma = \Gamma_d = g/c_p \qquad [3.7.12]$$

This is the dry adiabatic lapse rate defined previously, now denoted by the subscript d. With the known values $g = 9.807$ m s^{-2} and $c_p = 1.004 \times 10^3$ J kg^{-1} K^{-1}, the numerical result, to two significant figures, is

$$\Gamma_d = [9.807 \text{ m s}^{-2}]/[1.004 \times 10^3 \text{ J kg}^{-1} \text{ K}^{-1}]$$
$$= 9.8 \times 10^{-3} \text{ K m}^{-1} = 9.8 \text{ K km}^{-1}$$

where 1 J = 1 kg m^2 s^{-2}. This is the value stated in Section 3.6 without proof. A rising air parcel will cool by 9.8 K for each 1-km increase in altitude so long as water vapor is not condensing to liquid in the volume.

When the relative humidity in a rising air parcel increases to 100%, water vapor can begin to condense and latent heat is released. At this stage dq/dz becomes greater than zero, and its exact value depends on properties of the air parcel involved. If the parcel begins at the ground with a high humidity, r can reach 100% after rising only a short distance, and a large amount of latent heat can be released. However, if the parcel starts out with a low relative humidity, it has to rise further before it cools sufficiently for r to reach 100%. Latent heat release begins at different altitudes depending on details of the air parcel. Once dq/dz becomes positive, the temperature of the rising air parcel changes with a lapse rate described by Eq. 3.7.11. As the parcel continues to rise and cool, just enough water vapor keeps condensing to maintain the relative humidity in the volume at 100%. As temperature drops, the saturation vapor pressure decreases rapidly and water vapor continues to condense, keeping the actual water vapor pressure equal to the saturation value. After the parcel has experienced substantial cooling, there is insufficient water vapor left to cause further significant heating. At this point, the value of dq/dz becomes small, and if the parcel continues to rise, the lapse rate approximates the dry adiabatic value once again.

This derivation provides a conceptual model in which the lapse rate of the troposphere is the result of air parcels rising, cooling, and releasing latent heat, but what goes up must eventually come down. If all of the liquid water produced by condensation remained in the air parcel as it moved, then the processes already identified would occur in reverse when the parcel descends back to the ground. A descending parcel will warm, and if nothing else happens, the liquid water in it will evaporate and thereby consume heat from its surroundings. In this case, there will be no net transfer of energy to the atmosphere from the changes in phase of water. Heat is released to the atmosphere when a parcel rises and vapor condenses to form a cloud, but the same amount of energy is taken from the atmosphere when the liquid evaporates as the parcel descends. This reversible process does indeed occur in some cases, but it is not the entire picture. Once a cloud forms, new processes can come into play and produce rain or snow. Much of the liquid water falls out of the atmosphere as precipitation instead of remaining in the moving air parcels. Consequently, air parcels that descend from high in the troposphere reach the surface with very low relative humidities. They lose most of their water vapor on the ascending part of their journey, creating clouds and precipitation.

Figure 3.8 illustrates the atmospheric cycling of water. A parcel starts out warm at the ground and containing a water vapor amount corresponding to a relative humidity less than 100%. The parcel rises and cools until its relative humidity reaches $r = 100\%$ at the lifting condensation level, where a cloud forms. A significant amount of the liquid water falls out as rain under the action of gravity. The relative humidity in the rising parcel remains at 100%, but at the cold temperatures high in the troposphere, the absolute water vapor amount is extremely small. Eventually the parcel descends and warms. The relative humidity drops rapidly because of the increase in temperature and the accompanying increase in saturation vapor pressure. The parcel returns to the ground warm and very dry. This sequence implies that

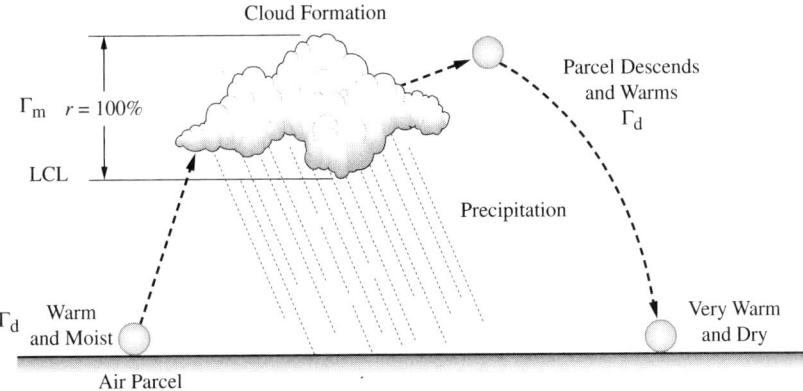

FIGURE 3.8   The cycling of atmospheric water. Water vapor condenses in rising air, and eventually falls out as precipitation. Precipitation ensures that latent heat release is not canceled by an equal consumption of energy in vaporization as parcels descend.

the transfer of latent heat from the ground to the atmosphere is a one-way process. Water vapor that starts out in an air parcel is eventually separated from this parcel and falls out as precipitation. The net result is the transfer of latent heat to the atmosphere.

The abundance of water vapor at the ground is known. Based on Table 1.2, a globally averaged mixing ratio is in the vicinity of 2×10^{-2} by volume. According to the theory developed here, if air parcels containing this amount of water vapor move up through the troposphere, the latent heat released in condensation leads to the observed globally averaged lapse rate of 6.5 K km^{-1}. This leads to the question, Is the observed water vapor abundance at the ground consistent with the observed lapse rate of the troposphere, all in a globally averaged sense? The theory says that these two quantities should be related. The dry adiabatic lapse rate, Eq. 3.7.12, should apply in regions of descending air or rising, unsaturated air, whereas Eq. 3.7.11 with $dq/dz > 0$ applies to ascending air where latent heat is released.

The combination of Eq. 3.7.11 and 3.7.12 gives $\Gamma = \Gamma_d - (1/c_p) \, dq/dz$, and rearrangement to determine dq yields $dq = c_p \, (\Gamma_d - \Gamma) \, dz$. Integration horizontally over the globe and from the ground to an altitude z^* gives the total latent heat released per unit mass of air. With the globally averaged value for Γ taken constant in altitude, this is

$$q = c_p \, (\Gamma_d - \Gamma) \, z^* \qquad [3.7.13]$$

The known values for use on the right-hand side of Eq. 3.7.13 are $\Gamma_d = 9.8$K km^{-1}, $\Gamma = 6.5$ K km^{-1}, and $c_p = 1.004 \times 10^3$ J kg^{-1} K^{-1}. A reasonable estimate for the depth of the troposphere over which latent heat is released is $z^* = 10$ km. Use of these values produces $q = 3.3 \times 10^4$ J kg^{-1}. This says that the observed lapse rate of the troposphere will result if 3.3×10^4 Joules of latent heat are released per kilogram of air over a 10-km vertical depth.

How much water vapor is required to release 3.3×10^4 J of latent heat per kilogram of air? This can be computed from Eq. 3.4.5. Integration over the depth of the troposphere produces

$$q = -\varepsilon \, L_c [\chi(z^*) - \chi(0)] \qquad [3.7.14]$$

where $\chi(z^*)$ is the water vapor mixing ratio at the apex and $\chi(0)$ refers to the ground. Temperatures high in the troposphere are very cold, so little water vapor can exist there. Therefore, it is acceptable to replace $\chi(z^*) - \chi(0)$ in Eq. 3.7.14 with $-\chi(0)$ alone. The result is

$$q = \varepsilon \, L_c \, \chi(0) \qquad [3.7.15]$$

Eq. 3.7.15 specifies the water vapor mixing ratio needed at the ground to provide a latent heat release q over the depth of the troposphere. If $q =$

3.3×10^4 J kg^{-1} of air as required by the observed lapse rate, then Eq. 3.7.15 produces a water vapor mixing ratio of 2.1×10^{-2}. This says that a globally averaged water vapor mixing ratio at the ground of about 2.1×10^{-2} is able to maintain the observed tropospheric lapse rate of 6.5K km^{-1}, and this value is consistent with the observed abundance. If the example uses a latent heat for converting vapor to ice, instead of condensing to liquid, the mixing ratio shrinks to 1.9×10^{-2}, which is also a reasonable result compared to the values in Table 1.2. Based on these numerical results, the theory developed to explain the temperature structure of the troposphere, expressed in Eq. 3.7.11, appears to be consistent with observations.

3.8 Vertical Motions, Stable Air, and Unstable Air

Rising air is a mechanism for the expansional cooling that increases relative humidity to the point of saturation. There are at least four ways to create vertical motions in the atmosphere. The first is thermal convection discussed in Chapter 1. Sunlight heats the Earth's surface, and dark-colored regions on the ground absorb more energy than do lighter-colored ones. Consequently, some parts of the ground in a given geographic region become warmer than others. A volume of warm air weighs less than the same volume of cooler air at the same pressure. The result is that parcels of air in contact with warmer areas on the ground can begin to rise, and this is the start of thermal convection. In summer, when heating of the ground is intense, convection is very efficient, and this can lead to cumulus clouds.

The second mechanism for creating rising motions is orographic lifting. This refers to the fact that air near the ground is forced to flow over the tops of objects that lie in its path. Mountain ranges are an important example of this process. Prevailing winds over North America move west to east, but along the way they encounter the Rocky Mountains and smaller obstacles. When air is forced upward, it cools and promotes the condensation of water vapor. The result is that the western slope of a mountain range experiences more cloudiness and rain than the eastern side (Lutgens and Tarbuck 2004). The eastern slopes tend to be clear and experience relatively little precipitation. The weather in Denver, located on the eastern side of the Rocky Mountains, is a good example of this effect.

Frontal activity is another common way to generate rising motions. On occasion, a relatively cold, dry mass of air moving into a region encounters a warmer, more humid air mass. The boundary between two air masses, with differing temperatures and water vapor contents, is called a "front." As the cold air mass advances, parcels of the warmer air are forced to rise upward along the front. This results in clouds and precipitation as the front moves by. Finally, as discussed in Chapter 4, low-pressure weather systems are regions of rising air, so clouds and precipitation tend to form here. The motions are a consequence of forces associated with horizontal differences in atmospheric pressure combined with the rotation of the Earth.

Assume that one of the identified mechanisms causes an air parcel to rise in altitude by a specific amount, and then the parcel is set free. Will the parcel sink back to the lower altitude where it started from, or will the parcel continue to rise on its own, without being forced by external means? The first case, where the parcel sinks, is called a "stable atmosphere." The second case, where the parcel keeps rising, is called an "unstable atmosphere."

Let the parcel begin its vertical motion at a level where the atmospheric pressure is P_1 and the temperature is T_1. The parcel then rises, its volume expanding along the way, to a level where the pressure is P_2. For simplicity, assume that the altitude separation between P_1 and P_2 is 1 km. It is necessary to distinguish between the temperature of the air parcel and the temperature of the surrounding atmosphere at the pressures P_1 and P_2. Conceptually, the parcel is a separate volume of air whose temperature is not always the same as that of the background atmosphere. At level P_1, let both the parcel and the surrounding atmosphere have the same temperature T_1. Also assume that the atmosphere's lapse rate is 6.5 K km^{-1}, so the atmospheric temperature at level P_2 is 6.5 K colder than at P_1. What is the temperature of the parcel when it reaches P_2? It is necessary to examine two cases.

In Case 1, water vapor is not condensing to liquid in the rising parcel or, equivalently, the relative humidity in the air parcel is less than 100%. If water vapor is not condensing, the temperature of the parcel changes at the dry adiabatic lapse rate of 9.8 K km^{-1}. After the parcel rises 1 km, its temperature is $T_1 - 9.8$ K. At the higher altitude, the parcel is colder than the surrounding atmosphere. The parcel has cooled by 9.8 K, but the atmosphere has cooled by only 6.5 K. The ideal gas law allows calculating atmospheric number density given pressure and temperature. From Eq. 1.8.1, $n = P/(kT)$, and this result applies both to the atmosphere and to the parcel at pressure P_2. The total number density of the atmosphere at P_2 is

$$n_{ATM} = P_2/[k(T_1 - 6.5 \text{ K})] \qquad [3.8.1]$$

while the number density of the parcel is

$$n_{PARCEL} = P_2/[k(T_1 - 9.8 \text{ K})] \qquad [3.8.2]$$

From the difference in temperature, it is obvious that the number density of the parcel is larger than the number density of the air, $n_{PARCEL} > n_{ATM}$. In general, colder air has a larger mass per unit volume than warmer air at the same pressure. The downward gravitational force on the parcel is greater than that on an equal volume of the surrounding atmosphere because the parcel is more massive. The result is that the air parcel will sink back to the pressure level that it started from. The parcel is akin to a rock thrown in a lake. A rock sinks because it is heavier than the same volume of surrounding water.

The described case is an example of "stable air" or a stable atmosphere. If the air parcel is forced to rise and then let go, it will simply sink back to the level where it began. The necessary condition for stable air is that a rising air parcel becomes denser than its surroundings. In response to this, it will sink. In practice, the parcel would not have started rising in the first place unless forced to do so.

In Case 1 the parcel had not yet reached its lifting condensation level. For Case 2, assume that water vapor is condensing to liquid in the parcel. An external force causes the parcel to rise, and after expansional cooling, the humidity in the parcel reaches 100%. Condensation begins, and latent heat is released in the air parcel as it rises from pressure P_1 up to P_2, a distance of 1 km. For the sake of illustration, assume that in response to latent heat release, the temperature of the parcel changes by only 5.0 K per kilometer of altitude. At pressure level P_2, the temperature of the ambient atmosphere is the same as in Case 1, $T_1 - 6.5$ K, but now the temperature of the parcel is $T_1 - 5.0$ K. As before, the ideal gas law gives the densities at pressure P_2. The number density of the atmosphere is as in Eq. 3.8.1, while the number density of the parcel is

$$n_{PARCEL} = P_2/[k(T_1 - 5.0 \text{ K})] \quad\quad\quad [3.8.3]$$

It is obvious that $n_{PARCEL} < n_{ATM}$. The parcel is warmer and therefore less massive than an identical volume of the surrounding air. Qualitatively, the parcel is like a helium-filled balloon; it is lighter than its surrounding, so it keeps rising on its own. In this case, once the parcel starts rising, it will continue to ascend on its own accord. There is no need to force it to move higher; it will ascend by merit of its own buoyancy until much of the water vapor has condensed. This is an example of "unstable air." Any temperature change for the air parcel less than 6.5 K km^{-1} leads to instability. The choice of 5.0 K km^{-1} in the preceding example is not unique.

The central ideas here are very simple. In stable air, a rising parcel becomes colder and denser than its surroundings, and therefore it will reverse direction and sink. In unstable air, a rising parcel becomes warmer and lighter than its surroundings, so it will continue to rise on its own. The examples used specific numerical values for the lapse rates to illustrate concepts, but the argument is more general. Let Γ_{ATM} be the lapse rate of the background atmosphere. The globally averaged value is 6.5 K km^{-1}, but in various geographic regions it can differ from this. Let Γ_{PARCEL} be the lapse rate of an isolated air parcel as it rises. Stable air corresponds to $\Gamma_{PARCEL} > \Gamma_{ATM}$, whereas unstable air requires $\Gamma_{PARCEL} < \Gamma_{ATM}$. In the context of this discussion, release of latent heat is the mechanism for creating instability. Notice, however, that release of latent heat is a necessary, but not a sufficient, condition for instability. On average, the lapse rate of the troposphere is $\Gamma_{ATM} = 6.5$ K km^{-1}, and this number already includes a contribution from latent heat release. The only way an air parcel can become unstable is for it to experience a relatively large release of latent heat, where "large" is defined by reference to conditions that prevail in the surrounding atmosphere. Unstable air parcels frequently exist on hot, humid days in summer. As these parcels ascend, a large quantity of water vapor condenses, and the rising motions can persist high into the troposphere. A visible consequence is the formation of tall cumulus clouds on summer afternoons in humid air masses. Hess (1959) and Bohren and Albrecht (1998) give additional information on atmospheric stability and instability.

The temperature inversion, introduced in Chapter 1, is an important special case of a stable atmosphere. In a temperature inversion, the temperature increases slightly in the first couple of kilometers above the ground, as shown in Figure 1.1. If an air parcel rose from the ground in this situation, it would immediately be colder and denser than its surroundings. Therefore, it would sink unless forced to continue rising. The atmosphere is extremely stable in a temperature inversion, and these conditions can be associated with air pollution episodes in urban areas, as the stability of the atmosphere inhibits the dispersal of pollutants by vertical mixing.

3.9 Condensation Nuclei and the Formation of Precipitation

The material presented thus far describes cloud formation as if it occurs spontaneously whenever the relative humidity reaches 100%. In practice, this is not the case. Water vapor seldom condenses to liquid or freezes to ice in completely clean air. The term "clean" here means air that is totally free of small particles. For water vapor to condense to liquid, molecules of water have to bond together, but individual molecules in the vapor phase seldom do this. Instead, they collide and fly off in another direction even when the relative humidity reaches 100%. This occurs because the attractive intermolecular forces are too weak to make rapidly moving gas phase molecules begin to clump together to form a liquid. If some liquid water is already present, then water in the gas phase can stick to it and condense, but the difficult part is getting the first bit of liquid. Consequently, the actual water vapor pressure in a volume of air can exceed the saturation vapor pressure, and condensation still does not occur. When relative humidity exceeds 100%, the air is "supersaturated."

Something in the atmosphere must initiate condensation, and small particles are important in this process. Small particles called "condensation nuclei" (CN) are the centers on which condensation from vapor to liquid begins in the atmosphere. Molecules of water vapor collide with and stick to these solid surfaces. The nuclei hold water molecules close together, attractive intermolecular forces come into play, and the first liquid appears here. A similar process operates when dew forms in the morning on grass or automobile windshields. At altitudes above the ground, small particles in the atmosphere provide the surfaces on which condensation begins.

The troposphere contains numerous small particles that act as CN. Smoke from forest fires is a natural source of CN, as is the breaking of ocean waves that introduce tiny salt particles into the atmosphere. In the modern world, industrial emissions are important. Smokestacks and engine exhaust account for a significant fraction of the particles in the present-day atmosphere, especially over land. The final source of atmospheric particles involves chemical reactions. Sulfur-containing gases released into the atmosphere participate in chemical reactions that eventually create small particles. Industrial activity is a major source of sulfur, but there are also natural sources. The oceans contain one-celled plants called phytoplankton, which engage in photosynthesis and serve as food for larger creatures. Phytoplankton release a molecule called dimethyl sulfide that leads to formation of small atmospheric particles that act as CN (Charlson et al. 1987).

As water vapor molecules collide with CN and stick to their surfaces, a coating of water gradually collects around the nuclei. Then, when the relative humidity in the air reaches or exceeds 100%, vapor will condense on the layer of water that surrounds the particle. Without a CN to hold them in place, the gas phase water molecules would collide, bounce off each other, and remain as vapor in spite of a relative humidity above 100%. The role of the condensation nucleus is to hold water molecules close together, thereby allowing attractive intermolecular forces to initiate the change in phase.

Table 3.3 lists some typical dimensions relevant to water vapor, condensation nuclei, and liquid water in the atmosphere. A water molecule has a radius on the order of 10^{-10} m, and a condensation nucleus has a radius 100 to 1000 times this, 10^{-8} to 10^{-7} m. When water condenses on condensation nuclei, cloud droplets form. A typical radius here is on the order of 10^{-5} to 10^{-4} m. The individual droplets in a cloud are in this size range, which is too small to see. At this point, a cloud has formed, but the issue of how to produce precipitation remains. A typical raindrop has a radius of order 10^{-3} m, 10 to 100 times larger than that of a cloud droplet (Goody and Walker 1972). Once cloud droplets form by condensation, some new processes become efficient, and these lead to the creation of raindrops. Two different mechanisms

TABLE 3.3 Characteristic Dimensions Relevant to Atmospheric Water and Clouds

- Radius of one water molecule $\sim 10^{-10}$ m
- Radius of a condensation nucleus $\sim 10^{-8}$ to 10^{7} m
- Radius of a cloud droplet $\sim 10^{-5}$ to 10^{-4} m
- Radius of a raindrop $\sim 10^{-3}$ m

can create raindrops starting from small cloud droplets. They are the "collision–coalescence process" and the "Bergeron process."

The collision–coalescence process operates in clouds where the temperature is above freezing, so water is present in the liquid state. When water vapor condenses to form cloud droplets, they tend to fall under the influence of gravity. However, these cloud droplets are not all the same size. Table 3.3 gives a range of radii from 10^{-5} to 10^{-4} m, and the large drops are more massive than the small ones. These larger, heavier droplets descend relative to the smaller, lighter ones because the updrafts that exist in clouds support only the smaller droplets. The light droplets are akin to ping pong balls being supported by the upward motions, while the bigger droplets are like baseballs that fall. A warm cloud consists of a few large droplets descending through a background of many smaller droplets. The big droplets collide with the small droplets, and in some cases, the small droplets coalesce with the larger ones. This is analogous to a snowball rolling downhill, accumulating the snow in its path and growing as it moves. The large drops eventually fall from the bottom of the cloud as raindrops. Wallace and Hobbs (1977) provide additional details of this process.

The described mechanism is conceptually straightforward; small droplets become incorporated into big droplets that grow still bigger in the process. Yet, coalescence is a difficult process to describe quantitatively. When a large droplet falls, air flows around its sides. This flow of air acts to divert the small droplets around the large ones, thereby inhibiting direct collisions. The same principle operates when a car moves rapidly along a highway in summer. Numerous insects are flying about at that time of year. If every bug headed for a car's windshield actually collided with it, the driver soon would be unable to see clearly. Fortunately, the flow of air over the windshield diverts most of the insects, and only a few unlucky ones get squashed. The same principle applies to the growth of raindrops by coalescence. In addition, the collision–coalescence mechanism assumes that some relatively large cloud droplets exist to begin with, and these would have to form in the original condensation process. In some cases, few of these large droplets are present, so the collision–coalescence process is not always an efficient way to create raindrops.

Much of the rain and snow received at middle latitudes is the result of the Bergeron process, named for the scientist who, with W. Findeisen, studied the mechanism during the 1930s. The discussion of cloud formation in Section 3.6 assumed that water vapor in a rising air parcel condenses to the liquid phase. Actually, many clouds reside at altitudes where the temperature is below freezing, so much of the condensed water exists as ice. The Bergeron process operates in clouds whose temperatures are below freezing. These cold clouds consist of a mixture of water in three phases: water vapor, ice crystals, and "supercooled" liquid. Observations show that ice particles and liquid droplets can coexist in clouds at temperatures below freezing. Section

3.3 established that the saturation vapor pressure over ice, $P_{SAT}(T)_{ICE}$, is smaller than the saturation vapor pressure over liquid, $P_{SAT}(T)_{LIQ}$, at the same temperature. That is, $P_{SAT}(T)_{ICE} < P_{SAT}(T)_{LIQ}$ where the temperature is less than 273 K. This inequality comes from the fact that the attractive intermolecular forces in ice are stronger than those in liquid water.

Consider the definition of relative humidity, $r = 100 \; P_{H_2O}/P_{SAT}(T)$, where P_{H_2O} is the true water vapor pressure. Based on the preceding discussion, does the saturation vapor pressure in the denominator of r refer to vapor over a liquid surface or over an ice surface? In a cold cloud, there are two different values for saturation vapor pressure, and it is necessary to distinguish between relative humidity with respect to a supercooled liquid droplet and relative humidity relative to an ice crystal. The same water vapor pressure leads to two different values of relative humidity. These are the relative humidity with respect to ice, $r_{ICE} = 100 \; P_{H_2O}/P_{SAT}(T)_{ICE}$, and the relative humidity with respect to liquid water, $r_{LIQ} = 100 \; P_{H_2O}/P_{SAT}(T)_{LIQ}$. In a cloud where $T < 273$ K, it is the case that $r_{ICE} > r_{LIQ}$; the relative humidity with respect to ice is greater than the relative humidity with respect to liquid water. This inequality is a major ingredient in the Bergeron process.

Imagine a situation in which the relative humidity with respect to ice exceeds 100%, but the relative humidity with respect to supercooled liquid water is less than 100%. The average of r_{ICE} and r_{LIQ} over the volume of the cloud might be 100%, but there will be supersaturation with respect to ice, while the air will be below saturation with respect to supercooled liquid water. Since $r_{ICE} > 100\%$, water vapor collects and freezes on the ice particles in the cloud, so the ice particles grow. However, because $r_{LIQ} < 100\%$, there is a net evaporation from the liquid. This evaporation supplies vapor that proceeds to freeze on the ice, so ice crystals grow at the expense of the supercooled liquid droplets. Observations show that ice crystals are less numerous than supercooled droplets. Because of their relative scarcity, the individual ice crystals can grow to the size where they precipitate out under the influence of gravity.

The Bergeron process operates because of differences in saturation vapor pressure over ice as compared to supercooled liquid water. These are the direct result of the different strengths of intermolecular forces in the liquid and solid phases, which trace back to the distribution of electrical charge in a water molecule. The texts by Wallace and Hobbs (1977) and Bohren and Albrecht (1998) contain additional information on processes that occur in clouds at temperatures below the freezing point of water.

When ice particles fall from the base of a cloud, they move through an atmosphere that becomes warmer as height decreases. If the air temperature near the ground is well above freezing, the ice is likely to melt during its fall, and the result is liquid rain. This is not always the case. Sometimes in summer, large clouds form from intense thermal convection, and the updrafts in these clouds are able to support large ice particles. The ice particles are carried upward by the strong vertical motions in the cloud and they continue to grow as they collide with smaller crystals and droplets along the way. Eventually the particles grow large enough that the updraft is unable to support them. At this stage, the particles descend, and a large ice particle falls from the base the cloud. The ice may be sufficiently large to reach the ground without completely melting. The result is hail, and on some of the hottest days in summer, ice falls from the sky.

3.10 The Global Water Cycle

All of the processes discussed in this chapter influence the cycling of water through the Earth's atmosphere and ultimately determine the amount of water that resides there. Figure 3.9 depicts the resulting "global water cycle" in which it is useful to distinguish between the portion of the atmosphere in contact with the oceans (the marine atmosphere) from that over land (the terrestrial atmosphere). Each box in Figure 3.9 denotes one of four reservoirs in which water resides, with an estimate of the mass of water in grams contained in each based on Gleick (1996). Arrows indicate the flux of water in g H_2O yr^{-1} that flows between any two reservoirs (Schlesinger 1997). A balanced global budget requires the flux that flows into each reservoir to equal the flux that exits that reservoir. A check of the values in Figure 3.9 shows that this is indeed the case.

Evaporation of water from the global oceans injects 425×10^{18} g H_2O yr^{-1} into the atmosphere, and 385×10^{18} g H_2O yr^{-1} returns to the oceans as precipitation. When integrated over all oceans on the planet, evaporative loss exceeds the gain from precipitation by 40×10^{18} g H_2O yr^{-1}, however, it

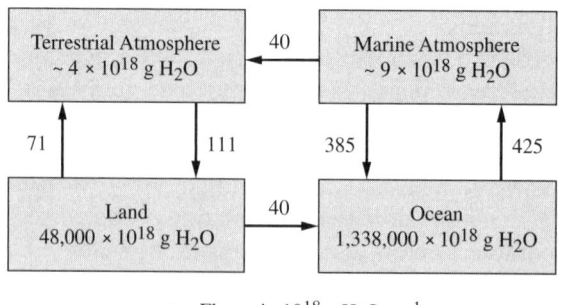

FIGURE 3.9 The global water cycle. Values in each reservoir give the mass of water, and arrows indicate fluxes between reservoirs. Over the oceans, the upward flux of water via evaporation exceeds the downward flux provided by precipitation. The opposite situation prevails over land.

is not the case that evaporation exceeds precipitation at all marine locations. As described by Wells (1997), exchange with the atmosphere leads to a net gain of liquid in the Pacific and Arctic Oceans, with part of the excess being removed by currents that move water into the Atlantic and Indian Oceans.

When averaged over all land surfaces, including ice, the gain in water by precipitation (111×10^{18} g H_2O yr^{-1}) exceeds the loss by evaporation and the release of vapor by plants, a process called transpiration. These terrestrial losses combine to give a flux of 71×10^{18} g H_2O yr^{-1}. The net effect is a gain of 40×10^{18} g H_2O yr^{-1} on land, which exactly offsets the net loss in water experienced by the global oceans. Transport by winds carries this excess from the marine atmosphere to the terrestrial atmosphere where it precipitates out. Finally, runoff by rivers and streams carries the surplus from the land back into the oceans.

A box diagram with fixed numerical values, as Figure 3.9 shows, provides a simplistic picture of the cycling of water. Several physical processes acting over different spatial scales create the fluxes that combine to produce an equilibrium state. Most of the global upward flux of water vapor provided by evaporation occurs over the large spatial extent of the oceans. The magnitude of this flux varies with water temperature, which itself responds to solar illumination and ocean currents. Over land the flux provided by evaporation varies with the geographic availability of surface water and soil moisture, while the contribution from transpiration depends on the distribution of vegetation.

Once water vapor enters the atmosphere it moves with the prevailing winds. Ascending motions provide the cooling required for air to become saturated and create the possibility of precipitation. These motions may take place over large geographic regions such as the tropics, on the regional scale of low-pressure weather systems (Section 4.3), or in highly localized convection cells that produce cumulus clouds (Section 3.6). Once a volume of air reaches saturation, sufficient condensation nuclei must be present to allow the formation of a cloud. Then on the scale of individual clouds, the formation of precipitation (Section 3.9) provides downward fluxes of water that combine over the globe to yield the values listed in Figure 3.9. The net transport of water vapor and droplets by winds from the marine to the terrestrial atmosphere has coevolved with the surface budget of liquid water that provides the runoff to the oceans required for a balanced budget.

A change in any portion of the global water cycle would trigger compensating changes elsewhere to reestablish a balanced state. For example, a warming of the ocean's surface would increase the evaporative flux of water vapor into the terrestrial atmosphere. This would be followed by an increase in precipitation, whose spatial pattern depends on details of the prevailing winds. Since water vapor is the most important greenhouse gas in the atmosphere, the capability to evaluate the global water cycle is a critical component of global climate models.

3.11 Exercises

1. The temperature at noon on a summer day is 303 K (85° to 86°F), and the relative humidity is 62%. To what value would the temperature have to drop for condensation of water vapor to liquid to begin? This is called the "dew point temperature."

2. Intense solar heating in the tropics, within 20° to 25° latitude of the equator, leads to rising air parcels in this region. As the parcels rise, assume that all of their water vapor condenses to liquid to form tropical clouds. Once in the upper troposphere, the air parcels flow north or south, and they eventually sink toward the ground in the extratropics, defined as the latitude band just outside of the tropics. Assume that the clouds formed remain in the tropics; they do not move with the air parcels to extratropical latitudes. Given this situation, explain qualitatively why the extratropics will be a region of clear skies and warm surface temperatures. As observational confirmation of this, the major deserts on Earth are located at extratropical latitudes.

3. Consider the situation described in Exercise 2. Let the subscript R denote air in the region of rising motions (the tropics) and let the subscript S refer to air in the region of sinking motion (the extratropics). The air temperature at the ground in the tropics is $T_R(z = 0 \text{ km}) = 305 \text{ K}$ (89°–90°F). The atmospheric surface pressure is $P = 1013$ mb, and the relative humidity at the ground is $r = 60\%$. Tropical air parcels rise from the ground to the altitude of $z = 8.0$ km. Assume that the tropical air flows north or south only in a thin layer located at this altitude, where air in both the tropics and the extratropics has the same temperature, $T_R(z = 8 \text{ km}) = T_S(z = 8 \text{ km})$. Given this, compute the temperature of the ground in the extratropics, $T_S(z = 0 \text{ km})$. Be sure to explain any assumptions made in reaching the answer. For those accustomed to English units, it would be useful to convert the temperature $T_S(z = 8 \text{ km})$ into degrees Fahrenheit to get an intuitive feel for the result.

4. If the answer to Exercise 3 is correct, the computed temperature at the ground in the extratropics is unreasonably high. Still, the fundamental idea that air rises in the tropics, loses water vapor by creating clouds, and then sinks in the extratropics is correct. A resolution to the difficulty posed by the result of Exercise 3 is to claim that a process, not included in that example, acts to cool the air as it moves. Cooling of air parcels by a net loss of longwave radiation is a potential way to resolve the problem. As an air parcel moves anywhere in the tropics and extratropics (upward, horizontally, and downward), assume that a net loss of longwave radiation produces a cooling rate of 2 K per day. This radiative cooling combines with the cooling and heating associated with vertical motions to yield a net overall temperature change that differs from that deduced in Exercise 3.

    The observed temperature at the ground in the extratropics is $T_S(z = 0) = 310$ K. Based on this information and the answer to Exercise 3, how long, in days, does it take for an air parcel to make the loop from the ground in the tropics, upward to 8.0 km, and eventually back down to the ground again in the extratropics?

5. Ice particles and supercooled water droplets exist in the same cloud at a temperature if $T = 260$ K. The water vapor pressure in the cloud is $P_{H_2O} = 2.05$ mb. Develop a quantitative argument to show that the liquid droplets will tend to evaporate and the ice particles will tend to grow.

3.12 References

Bohren, C. F., and B. A. Albrecht. 1998. *Atmospheric Thermodynamics.* New York: Oxford University Press.

Botkin, D. B., and E. A. Keller. 1998. *Environmental Science. Earth as a Living Planet.* New York: John Wiley and Sons.

Charlson, R. J., J. E. Lovelock, M. O. Andreae, and S. G. Warren. 1987. Oceanic phytoplankton, atmospheric sulfur, cloud albedo and climate. *Nature.* 326:655–61.

Fermi, E. 1956. *Thermodynamics.* New York: Dover Publications.

Gleick, P. H. 1996. Water Resources. In *Encyclopedia of Climate and Weather,* S. H. Schneider, ed. New York: Oxford University Press.

Goody, R. M., and J. C. G. Walker. 1972. *Atmospheres.* Englewood Cliffs, N.J.: Prentice–Hall.

Hecht, E. 1996. *Physics: Calculus.* Pacific Grove, Calif.: Brooks Cole Publishing Co.

Hess, S. L. 1959. *Introduction to Theoretical Meteorology.* New York: Holt, Rinehart and Winston.

Lutgens, F. K., and E. J. Tarbuck. 2004. *The Atmosphere—An Introduction to Meteorology.* Upper Saddle River, N.J.: Pearson Prentice–Hall.

Mason, B. J. 1957. *The Physics of Clouds.* Oxford: Clarendon Press.

Penner, J. E., D. H. Lister, D. J. Griggs, D. J. Dokken, and M. McFarland, eds. 1999. *Aviation and the Global Atmosphere.* Cambridge, U.K.: Cambridge University Press.

Planck, M. 1945. *Treatise on Thermodynamics.* Translated by A. Ogg. New York: Dover Publications.

Rogers, R. R., and M. K. Yau. 1989. *A Short Course in Cloud Physics.* 3rd ed. Woburn, Mass.: Butterworth-Heinemann.

Schlesinger, W. H. 1997. *Biogeochemistry: An Analysis of Global Change.* 2nd ed. San Diego: Academic Press.

U.S. Standard Atmosphere, 1976. Washington, D.C.: U.S. Government Printing Office, 1976.

Wallace, J. M., and P. V. Hobbs. 1977. *Atmospheric Science—An Introductory Survey.* New York: Academic Press.

Wells, N. 1997. *The Atmosphere and Ocean.* New York: John Wiley and Sons.

Chapter 4

Winds: The Global Circulation and Weather Systems

INTRODUCTION

Motions in the atmosphere span spatial scales from local to global. Prominent among these are the general circulation, which encompasses the entire planet, and weather systems, which influence length scales on the order of several hundred kilometers. Horizontal winds originate when gradients in pressure act to accelerate air parcels. The rotation of the Earth then produces a Coriolis effect that deflects an air parcel to the right of its instantaneous direction of motion as viewed from a fixed point on the surface of the planet in the Northern Hemisphere. An equilibrium wind is possible because the Coriolis acceleration depends on wind speed, whereas the pressure gradient force is independent of air motion. When air parcels move in a curved path, the centrifugal force is an added complication, and frictional forces can be significant near the ground.

A balance among several forces leads to a global scale circulation in which winds in the middle troposphere move from west to east in both hemispheres. On smaller spatial scales, low-pressure weather systems are characterized by horizontal surface winds that flow counterclockwise about a core of minimum pressure, plus an inward component that provides for ascending motions with the formation of clouds and precipitation. High-pressure systems are the opposite case and display horizontal winds that flow clockwise and outward, with descending vertical motions and clear skies. An overall effect of the winds associated with low- and high-pressure systems is the transport of heat toward high latitudes, a process that compensates for the net radiative cooling experienced in these regions.

4.1 Wind Systems in the Earth's Atmosphere

Viewed from a fixed location in outer space, the atmosphere is a spherical gaseous shell sitting atop a rotating planet. A point on the Earth's surface moves around in a circle as the planet rotates, and the speed associated with this rotation varies with latitude. At the equator, the west-to-east speed is approximately 4.6×10^2 m s^{-1}, decreasing to 4.0×10^2 m s^{-1} at 30° latitude, to 2.3×10^2 m s^{-1} at 60°, and ultimately to 0 m s^{-1} at either pole.

One can envision two extreme scenarios for the relative motion between the atmosphere and the Earth. In one case, the interface between the atmosphere and the solid Earth is "slippery," and the atmosphere remains stationary, unaffected by the rotating planet beneath it. Here an observer hooked to the planet's surface would experience very strong winds moving from east to west and equal in magnitude to the Earth's rotation speed at that latitude. The inhabitants of this hypothetical planet would live with perpetual winds far in excess of those encountered in the strongest hurricane or tornado. In the other extreme, frictional forces exist between the atmosphere and the solid Earth, and as a consequence the atmosphere is dragged along with the rotation of the far more massive planet. In this case the inhabitants experience no winds at all. That winds measured from the surface of the Earth are typically several m s^{-1} up to several tens of m s^{-1} indicate that the latter case is much closer to reality than the former.

At any given location and time, the wind measured by an observer on the Earth's surface might appear to move in an erratic, seemingly random manner, but viewed in a time-averaged sense, several different types of organized wind systems exist in the atmosphere. These systems differ from each other by their characteristic horizontal dimensions. The "global circulation," also called the "general circulation," represents the largest spatial scale. The global circulation is a well-defined wind system that moves around the entire planet. The characteristic spatial scale is planetary in size, on the order of 10^4 km. As part of the global circulation, an air parcel might travel from the equator to the North Pole, and along the way, it can move around the world from west to east several times. One objective of this chapter is to identify the forces responsible for driving this circulation and to explain how they combine to produce the observed winds. Table 4.1 summarizes some characteristics of the global circulation as well as the two remaining classes of wind system described in the following.

TABLE 4.1   Classes of Wind Systems in the Earth's Atmosphere

1. Global Circulation (General Circulation)
Air moves around the entire planet in an organized way.
Spatial scale $\sim 10^4$ km (planetary scale).
2. Weather Systems (Low- and High-Pressure Systems)
Air circulates around regions of relatively low or high surface pressure.
Spatial scale $\sim 10^2$–10^3 km (mesoscale).
Systems move about, embedded in the global circulation.
3. Local Circulations
Arise from local conditions (land–water contrasts, mountains, air masses with contrasting properties).
Spatial scale $\sim 10^1$ km (local).

On a smaller spatial scale, the atmosphere contains "weather systems." These are regions of relatively high or low atmospheric pressure as measured at the ground, and they have characteristic dimensions of several hundred to a thousand kilometers. A classic problem in theoretical meteorology involves trying to explain how these weather systems develop, particularly the low-pressure areas. Weather forecasts that extend more than several days into the future tend to become inaccurate, and one of the reasons is a limited ability to predict the development of new low-pressure systems and their evolution over time. The left side of Figure 4.1 illustrates the appearance of an idealized low-pressure system on a weather map. The contours, called "isobars," are lines of constant atmospheric pressure at the ground. Isobars need not be circular as in Figure 4.1, but the isobars of a low-pressure system will consist of closed loops. Given a typical sea-level pressure of 1013 mb, the minimal central pressure of 990 mb depicted in Figure 4.1 would be a very deep low, but not a record breaker. A hurricane is an extreme case of a region of low pressure. For example, while over the Gulf of Mexico in September 2005, Hurricane Rita had a minimum pressure near 900 mb. A high, depicted on the right side of Figure 4.1, displays the opposite pattern, in which pressure at the Earth's surface increases toward the center of the system. These regions of relatively low and high surface pressure move about embedded in the global circulation. Low- and high-pressure weather systems have characteristic wind patterns associated with them, and a major topic of this chapter centers on determining the mechanisms responsible for these motions.

The final category of wind system consists of a variety of "local circulations." These arise from highly localized conditions, and the typical spatial scales are on the order of 10^1 km. For example, temperature contrasts be-

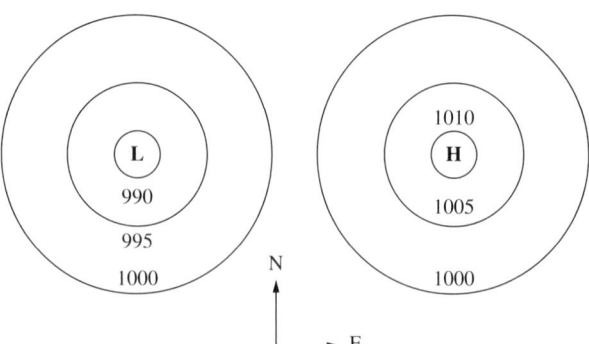

FIGURE 4.1   Idealized structure of low-pressure (left) and high-pressure (right) weather systems viewed from above. Atmospheric pressure at the ground is constant along closed lines called isobars. Isobars are labeled in millibars, where 1000 mb is a typical surface pressure.

tween land and water can lead to characteristic wind patterns confined near a shoreline. Proximity to mountains can produce local winds; locations on the front range of the Rocky Mountains occasionally experience very strong winds associated with diversion of air over and around these natural barriers. Finally, a tornado represents an extreme case of a localized circulation that can develop when air masses with different temperatures and water vapor contents encounter each other.

4.2 Forces that Drive the Winds

Why does air start moving in the first place? For a parcel of air to start from rest and accelerate, it must experience a nonzero net force. Newton's Laws of Motion describe the relationship between forces and motion. Newton's First Law says that an object continues in a state of rest or motion at constant velocity unless acted on by a nonzero net force (e.g., Hecht, 1996). The object of interest in this chapter is a parcel of air. Several different forces can act simultaneously, and when this occurs the object responds to the sum of the individual forces. This leads to Newton's Second Law, which says that when a nonzero net force acts on an object, the object accelerates in the direction of this net force, and acceleration is proportional to force. As described in Chapter 1, the quantitative statement of the Second Law is Force = Mass × Acceleration, where for the study of winds, it is important to recognize that force and acceleration have both magnitude and direction.

Wind systems in the Earth's atmosphere are a response to several different forces acting simultaneously. In this regard, it is useful to consider a simple case to show what happens when more than one force acts on an object. This example, which involves forces experienced by a falling object, does not refer to atmospheric winds, but it illustrates an important concept that will prove relevant in subsequent work. If an object falls from the top of a tall building, it will accelerate downward. As discussed in Chapter 1, for every second that goes by, the gravitational pull of the Earth acts to increase the downward velocity. The force of gravity is

$$F_{\text{down}} = -mg \qquad [4.2.1]$$

where m is the mass of object, $g = 9.8$ m s^{-2} is the acceleration due to gravity, and the minus sign indicates that the force is downward. If gravity were the only force that acted on the object, it would continue to accelerate. The object's velocity would keep increasing until it hit the ground.

Gravity is not the only force that acts on an object falling through the atmosphere. There is a second force called "air resistance." The object experiences air resistance as it falls because it has to push air molecules out of its way. Air resistance is an upward force that opposes motion, and both forces act on the object simultaneously. The force associated with air resistance depends on the velocity of the falling object. For purposes of illustration it is sufficient to assume that the upward force, F_{up}, is proportional to the vertical velocity, w:

$$F_{up} = +\beta w \qquad [4.2.2]$$

The plus sign denotes an upward force, the constant β depends on the shape of the object and the density of atmosphere, and w is positive in the downward direction. Eq. 4.2.2 shows that as the fall velocity increases, the upward force becomes larger. In contrast to this, the downward force of gravity does not depend on velocity. The combination of Newton's Second Law with expressions for the two forces produces an equation that describes the velocity of the object as a function of time t:

$$m \, dw/dt = mg - \beta w \qquad [4.2.3]$$

where dw/dt is the vertical acceleration of the mass m. The velocity of the object will increase up to the value at which the upward force provided by air resistance exactly balances the downward force of gravity. At this stage, the object stops accelerating, $dw/dt = 0$, and the balance of forces requires

$$mg - \beta w = 0 \qquad [4.2.4]$$

Eq. 4.2.4 specifies an equilibrium velocity or terminal velocity, called $w*$, where

$$w* = mg/\beta \qquad [4.2.5]$$

According to Newton's First Law, when the net force becomes zero, an object moves at a constant velocity from that point on. Once the object reaches this special velocity, it no longer accelerates, and it keeps falling at this constant speed all the way to the ground.

This example illustrates motion under the condition of balanced forces. The term "balanced forces" simply means that the net force is zero. A balance among several forces is possible because one or more of the forces depends on velocity. In the preceding example, this is air resistance. The velocity of the object is an adjustable parameter that varies until it reaches the specific value where the net force is zero. When this is the case, acceleration is also zero, and the velocity remains constant. The same type of reasoning can be applied to the motions of air parcels in the atmosphere. The observed wind systems represent motions of air parcels under balanced forces, at least in a temporally averaged sense. The forces that drive winds in the atmosphere are different from the forces considered in the example, but the principle is the same. Some of the forces that drive the wind depend on the velocity of an air parcel, and some do not. Under the action of several forces, the speed of an air parcel in the atmosphere adjusts until it experiences a net force of zero. When this balance is reached, the air continues to move at a constant speed.

The following sections describe four forces that are responsible for producing the wind systems observed in the Earth's atmosphere. These are the pressure gradient force, the Coriolis force, the centrifugal force, and friction. The pressure gradient force is responsible for initiating the motion of air. Once an air parcel starts moving, other forces that depend on the wind speed come into play. In response, the wind speed adjusts until all forces acting together sum to zero. The general circulation and the motions around high- and low-pressure weather systems arise from the simultaneous action of these forces.

4.3 The Pressure Gradient Force

The pressure gradient force may be the most fundamental of the four in that it initiates the motion of air, and it does not depend on the air's speed. The pressure gradient force expresses the intuitive fact that air tries to move from regions of high pressure to regions of low pressure. A simple example that uses liquids instead of gases illustrates this point (Goody and Walker 1972). Figure 4.2 depicts two identical containers that are open at the top. These containers are connected at the bottom by a thin tube with a valve that can open and close. Let Δy be the distance between the containers, where, to avoid any ambiguity in defining its value, Δy is very large compared to the diameter of the containers. Initially the valve is closed, and container 2 holds

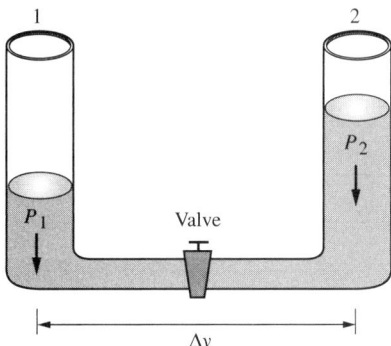

FIGURE 4.2   A difference in pressure over a distance Δy leads to acceleration of liquid from a region of high pressure to a region of low pressure. In equilibrium, the pressure gradient is zero.

a large amount of liquid, whereas only a small quantity exists in container 1. The downward pressure in container 2, P_2, is greater than that in container 1, P_1. Over the horizontal distance Δy there is a difference in pressure equal to $\Delta P = P_2 - P_1$. The "pressure gradient" is defined as the difference in pressure divided by the distance over which this pressure difference exists:

$$\text{Pressure Gradient} = (P_2 - P_1)/\Delta y = \Delta P/\Delta y \qquad [4.3.1]$$

When the valve that connects the containers is opened, liquid flows from container 2 to container 1 until the same amount of liquid exists in both containers. In this final state, $P_2 = P_1$, or the pressure gradient is zero, $\Delta P/\Delta y = 0$. The key idea is, Whenever a pressure gradient exists in a fluid, motions develop that act to eliminate this pressure gradient. This applies to a liquid or to a gas like the atmosphere. A "pressure gradient force" acts on the fluid, and this force leads to motions that attempt to eradicate the pressure gradient, unless opposed by other forces. The force always points from regions of high pressure, container 2, to regions of low pressure, container 1.

This reasoning applies to motions in the Earth's atmosphere. As a simple illustration, assume that the atmosphere's temperature at any altitude is the same at all latitudes from the equator to the polar regions. Figure 4.3 depicts a cross section of atmospheric pressure in altitude and latitude for this special case. In the absence of horizontal variations in temperature, surfaces of constant atmospheric pressure are concentric circles around a spherical planet at constant altitude. The atmospheric pressure at the ground is about 1000 mb. Somewhere above the ground, there is a surface where the pressure is 900 mb, and so on to higher altitudes and lower pressures. There is no horizontal pressure gradient in this idealized atmosphere and hence no horizontal pressure gradient force.

Note that a vertical pressure gradient always exists in the atmosphere because the pressure at the ground is greater than the pressure at a higher altitude. A vertical pressure gradient force always tries to push air upward

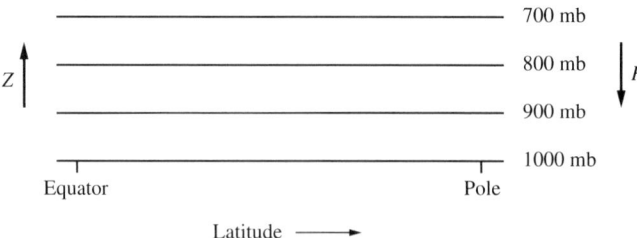

FIGURE 4.3   An idealized atmosphere in which the downward pressure P at any fixed altitude z is independent of latitude. No horizontal pressure gradient force exists in this case.

from high pressure to low pressure. Yet, if this is true, why doesn't the atmosphere fly away into outer space? The answer is that the downward force of gravity balances the upward pressure gradient force, and this is the state of "hydrostatic balance" defined in Chapter 1. Neither the pressure gradient force nor gravity depends on the air speed, so this balance does not involve the generation of winds.

The case depicted in Figure 4.3 has no horizontal pressure gradients. At any fixed altitude, the pressure is the same at all latitudes. In this artificial scenario, there is no horizontal pressure gradient force and no horizontal wind. The true atmosphere differs from Figure 4.3 in that regions near the equator are warmer than those near the poles. The warm air near the equator tends to expand upward, since a given mass of warm air occupies a greater volume than the same mass of colder air at the same pressure. As a result of temperature differences between the equator and poles, lines of constant pressure are not horizontal. Instead, they slant upward from the pole to the equator, as illustrated in Figure 4.4, where the magnitude of the slope is grossly exaggerated for purposes of illustration. In this circumstance, a horizontal pressure gradient exists everywhere in the atmosphere except at the ground. If A-B is a horizontal line, then the pressure at point A is greater than the pressure at point B. This inequality applies to altitudes above the ground and in particular to the middle and upper troposphere. An air parcel placed at point A experiences a pressure gradient force that pushes it toward point B, from high pressure to low pressure. The concepts here are straightforward: A horizontal pressure gradient exists in the middle to upper troposphere because temperature varies between the equator and pole. This pressure gradient force causes a parcel of air to accelerate from low latitudes to high latitudes.

When a mass of air flows away from a point, it is necessary for an equal mass to flow toward the point so as not to leave behind a vacuum. This concept is called "continuity of mass," and it is an important constraint on the

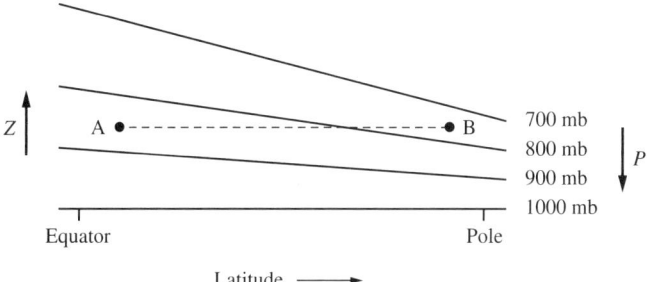

FIGURE 4.4 An atmosphere in which the equatorial regions are warmer than the polar regions. Lines of constant pressure P slope downward in altitude Z from equator to the pole, leading to a horizontal pressure gradient force that points from point A to point B.

motion of a fluid. The tropical regions experience intense heating of the ground by sunlight, and this drives vertical convection. The upwelling continually supplies air parcels that then move toward the pole under the influence of the pressure gradient force. Figure 4.5 illustrates the resulting flow pattern. Air eventually descends over the polar regions. This descending air appears to flow counter to the upward pressure gradient force, but recall that gravity offsets this. Finally, as required by continuity of mass, air at low altitudes flows back toward the tropics. The resulting cellular pattern constitutes a simple equator-to-pole circulation. This is called a "Hadley cell" or "Hadley circulation" after George Hadley, who proposed it in 1735 (Lorenz 1967). The actual global circulation is much more complicated than is depicted in Figure 4.5, but the fundamental idea that the pressure gradient drives winds between the equator and poles is correct.

Figure 4.6 illustrates the form of the Hadley circulation on a spherical Earth. Air rises in the equatorial regions, followed by a flow from low to high latitudes. Air converges toward the poles and sinks, making the return flow toward the equator at low altitudes. The Hadley circulation considers only vertical motions and flow between the equator and poles. It says nothing about winds that move from west to east. The Hadley cell alone cannot possibly provide a complete explanation of the global circulation for at least two reasons. First, to achieve an equilibrium circulation, the pressure gradient force must be balanced by at least one additional force whose magnitude depends on wind speed. Second, the Hadley circulation takes no account of the fact that the Earth is rotating, and as will be shown in Section 4.4, this is a major influence on the circulation of the atmosphere.

The pressure gradient force is also important in determining winds associated with high- and low-pressure weather systems. As described previously, lines of constant atmospheric pressure at the ground form closed contours around centers of minimum or maximum pressure. Figure 4.7 illustrates a low- and a high-pressure system as viewed from above. The observed

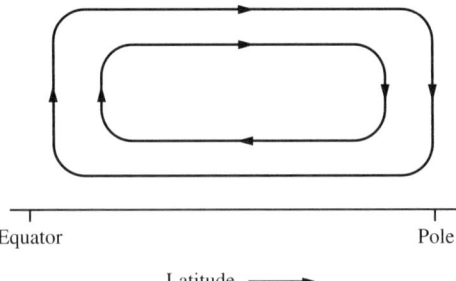

Equator   Pole

Latitude ⟶

FIGURE 4.5 The Hadley circulation consists of rising air over equatorial latitudes, flow toward the pole in the middle and upper troposphere, sinking air at high latitudes, and flow back toward low latitudes near the ground.

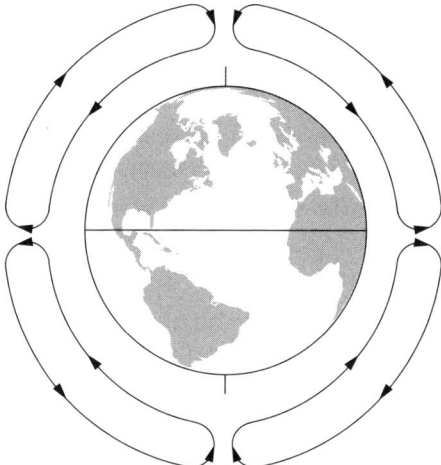

FIGURE 4.6   Schematic of the Hadley circulation on a spherical planet. The cellular pattern leads to a net transport of heat from tropical latitudes to the polar regions.

pattern of winds around these systems represents the combined action of all four forces identified earlier. However, the arrows in Figure 4.7 refer to wind directions at the ground under the unrealistic assumption that the pressure gradient force acts alone. The force generates a flow of air from high pressure to low pressure, so air near the ground flows inward toward the center of a low and outward from the center of a high. The actions of friction, the centrifugal force, and especially the Coriolis force modify this simple description considerably, but it is still true that winds associated with a low have

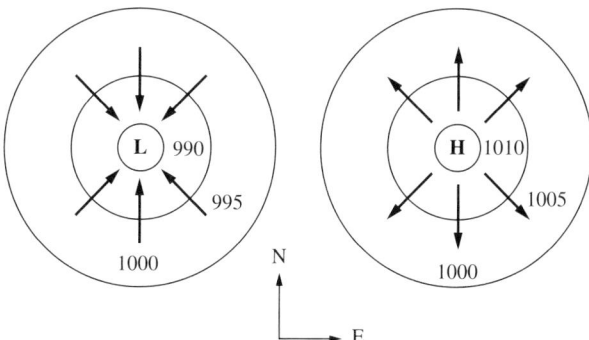

FIGURE 4.7   Horizontal motion of air associated with low-pressure (left side) and high-pressure (right side) weather systems under the assumption that only the pressure gradient force operates. An equilibrium circulation is not possible in this hypothetical scenario.

a component inward toward the center, whereas winds associated with a high have an outward component.

Figure 4.8 considers vertical motions that develop in low- and high-pressure systems. Based on the pressure gradient force alone, air at the ground would flow horizontally inward toward the center of a low from all sides. When air from different directions meets in the middle, the only way to go is up. Continuity of mass requires air to rise in response to the other air piling in behind it, so vertical motions in a low-pressure region are upward. As described in Chapter 3, when air rises, it cools, and the relative humidity increases. Eventually the relative humidity may reach 100%, and clouds form. Because of the upward motion, low-pressure regions tend to be cloudy, and much of the precipitation received at middle latitudes is associated with the passage of these systems. A region of "convergence" is one where air flows inward from all directions toward a point. Lows are regions of convergence at the ground where air in the central area ascends. Notice that air flowing inward effectively acts to fill in the region of low pressure; the flow of air in response to a pressure gradient tends to eliminate that pressure gradient. Based solely on this reasoning, low-pressure weather systems must eventually self-destruct, although as described later, other forces reduce the magnitude of the inward flow.

Motions in a high-pressure system are the opposite of those just described. Figure 4.7 shows that air flows horizontally outward at the ground from the center of a high, and this constitutes a "divergence" of air. The outflow must be supplied by descending motions near the center, as illustrated in Figure 4.8. When air descends, it warms and the relative humidity shrinks.

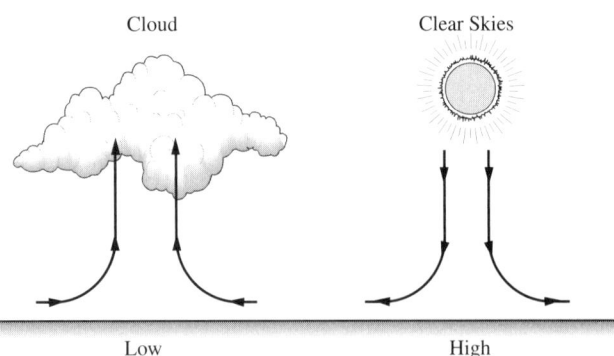

FIGURE 4.8   Vertical flow of air associated with low-pressure (left) and high-pressure (right) weather systems in the hypothetical circumstance that only the pressure gradient force operates. An equilibrium circulation is not possible when only one force acts.

Clouds cannot form in these conditions, so high-pressure systems are accompanied by clear skies. Sometimes high-pressure regions are called "fair weather systems."

The true wind patterns associated with atmospheric low- and high-pressure systems are more complicated than indicated in Figures 4.7 and 4.8. However, the conclusions reached about surface convergence with rising air in lows and divergence with sinking air in highs are correct, although other forces lead to components of motion in addition to those identified thus far.

It is possible to derive a mathematical expression to show how a parcel of air responds when subjected to a pressure gradient. Imagine a volume of air in the shape of a cube sitting at an arbitrary location, labeled by coordinates x, y, z, as shown in Figure 4.9. The dimensions of the cube are Δx by Δy in the horizontal and Δz in the vertical. Assume there is a pressure gradient in the $+y$ direction, so pressure increases from left to right in Figure 4.9. Air molecules are striking the left face of the cube, and this exerts a pressure, P, in the $+y$ direction. Molecules of air are also striking the right face of the cube. Let this pressure be $-(P + \Delta P)$, where $\Delta P > 0$ is the change in air pressure across the distance Δy, and the minus sign indicates that this pressure is exerted in the $-y$ direction.

The definition of pressure as the force per unit area provides an expression for the net force F_y on the cube in the $+y$ direction:

$$F_y = P\,\Delta x\,\Delta z - (P + \Delta P)\,\Delta x\,\Delta z \qquad [4.3.2]$$

The force is negative when ΔP is positive, or the force points from high pressure to low pressure, consistent with physical reasoning. Eq. 4.3.2 specifies the force for use in Newton's Second Law.

$$M\,a_y = M\,dv/dt = F_y \qquad [4.3.3]$$

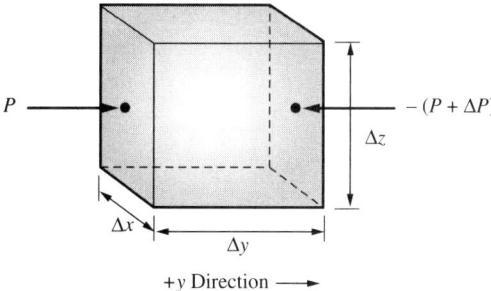

+y Direction ⟶

FIGURE 4.9   A volume subjected to a pressure gradient force. A pressure P directed toward the right acts on the left face, and a pressure $P + \Delta P$ directed toward the left acts on the right face. When $\Delta P > 0$, the cube accelerates to the left.

where a_y is acceleration and v is velocity in the $+y$ direction. The mass M of interest here is the mass of the cube. If ρ is the mass density of air, then

$$M = \rho \, \Delta x \, \Delta y \, \Delta z \qquad [4.3.4]$$

The combination of Eqs. 4.3.2, 4.3.3, and 4.3.4 gives

$$\rho \, \Delta x \, \Delta y \, \Delta z \, dv/dt = P \, \Delta x \, \Delta z - (P + \Delta P) \, \Delta x \, \Delta z \qquad [4.3.5]$$

or

$$dv/dt = - (1/\rho) \, \Delta P/\Delta y \qquad [4.3.6]$$

When the changes in pressure and distance are infinitesimals, $\Delta P/\Delta y$ can be replaced by a derivative:

$$a_y = dv/dt = -(1/\rho) \, dP/dy \qquad [4.3.6]$$

Eq. 4.3.6 says that the acceleration of the volume of air is proportional to the pressure gradient and in the opposite direction. If there is a pressure gradient along the $+x$ direction, analogous reasoning yields

$$a_x = du/dt = -(1/\rho) \, dP/dx \qquad [4.3.7]$$

where u is the velocity of the air parcel along the x axis.

A pressure gradient could exist in the z direction as well, but the downward force of gravity is important here. As stated previously, the downward force of gravity balances the upward pressure gradient force throughout most of the atmosphere, and the vertical acceleration of air is therefore zero. However, in limited geographic regions, such as in a thunderstorm, it is possible to have a local imbalance between the pressure gradient and gravitational forces. In such cases, complex, time-dependent motions can occur.

Eqs. 4.3.6 and 4.3.7 represent first steps toward describing horizontal winds in the atmosphere, but thus far only the pressure gradient force has entered the derivation. To achieve a state of balanced forces, it is essential to consider the Coriolis force. When a term representing the Coriolis force is added onto the right-hand sides of Eqs. 4.3.6 and 4.3.7, the result is a reasonably good mathematical model of atmospheric winds.

4.4 The Coriolis Force

The Coriolis effect is connected to the rotation of the Earth on its axis. People sitting on the Earth are usually unaware that they are moving. As pointed out in Section 4.1, however, all humans and objects are whirling around in a big circle as they move along with the Earth's rotation. The concept of acceleration includes both a change in speed and a change in the direction of motion over time. Hence, rotation is a form of acceleration, and this acceleration needs to be included in Newton's Second Law when applied to explain air motions observed in a reference frame that is spinning (e.g., Hecht 1996). The Coriolis effect acts as an acceleration or a force, although unlike gravity, for example, it is not a fundamental force of nature. Rather, the "Coriolis acceleration" and the "Coriolis force" arise because winds are measured relative to the surface of a rotating planet.

It is useful to describe what the Coriolis force does before attempting to explain why it behaves this way. If a parcel of air is moving in the Northern Hemisphere, the Coriolis force acts to bend the path of the parcel to the right of its direction of motion. The Coriolis force always points to the right of the instantaneous direction of motion, and its magnitude depends on velocity measured with respect to the ground. It does not matter in what direction an air parcel is moving; the Coriolis force points to the right of that direction in the Northern Hemisphere. In the Southern Hemisphere the Coriolis force always bends the path of air to the left. Because of the Coriolis force, air does not simply move from high pressure to low pressure as the pressure gradient force says it should. As described in Section 4.3, the pressure gradient force acting alone would make an air parcel move from equator to pole in the middle and upper troposphere of either hemisphere. Suppose an air parcel begins to move northward from low latitudes toward the North Pole. As soon as the parcel moves, the Coriolis force comes into play and deflects the motion toward the right, which is to the east, of its instantaneous velocity. If the poleward motion was in the Southern Hemisphere, the Coriolis force would point to the left, which also corresponds to eastward. The analysis presented in Section 4.7 shows that the combination of the pressure gradient and Coriolis forces leads to a global circulation, where the strongest winds move from west to east in both hemispheres. This motion is perpendicular to expectations based on the pressure gradient force alone.

Why does the Coriolis force behave as described? Instead of considering a spherical planet, it is easier to think about motions on a flat rotating turntable, as depicted in Figure 4.10 (e.g., Goody and Walker 1972). When viewed from above, the table is rotating counterclockwise, so this is akin to looking down on the Northern Hemisphere from a stationary point located above the North Pole. To make the situation analogous to the Earth, let the turntable complete one full rotation in a time $\kappa = 24$ h. Consider two different circles drawn on this turntable, one with radius r_A and the other with radius r_B, as shown in Figure 4.10. Points A and B are specific points on these circles, and both points complete a full rotation in the time κ. It is straightforward to compute the speeds of points A and B as measured by an

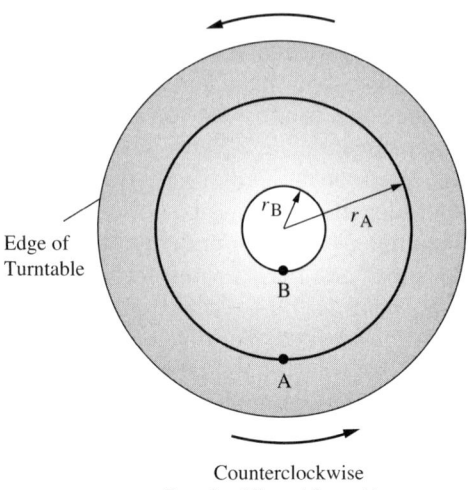

Counterclockwise
Rotation Viewed from Above

FIGURE 4.10   A rotating turntable illustrates the Coriolis force. A point located on the circle of radius r_A has a greater absolute speed than a point at radius r_B. When an object moves radially inward, its path turns to the right.

observer who is not rotating with the turntable. Point A moves a distance $2\pi r_A$ around the circle in time κ, so the speed of point A is

$$U_A = 2\pi r_A/\kappa$$

while the speed of point B is

$$U_B = 2\pi r_B/\kappa$$

The rotation rate of the turntable, in radians per unit time, is defined as $\Omega = 2\pi/\kappa$. Given this, the speeds of points A and B, hooked to the turntable, are

$$U_A = \Omega r_A \text{ and } U_B = \Omega r_B \qquad [4.4.1]$$

It is obvious that $U_A > U_B$. These speeds in a circle would be measured by an onlooker, called observer 1, who is located above the turntable and is not rotating. However, consider the situation as viewed by someone, called observer 2, sitting at point B, rotating with the turntable. Observer 2 would claim that both point A and point B are at rest, a conclusion that appears reasonable because points A and B remain the same distance apart. However, this observation is an artifact that arises because observer 2 is moving in a circle along with points A and B. In fact, observer 1, looking down from above, knows that points A and B are moving in a circle, that they have different speeds, and that speed increases with increasing distance from the center of the turntable. People on a planet are in the position of observer 2; the points A and B appear to be at rest when viewed from this rotating, accelerating frame of reference.

Consider an air parcel sitting at point A and rotating with the turntable. Measured relative to point A, the parcel is at rest, but as seen by observer 1 above the turntable the parcel is moving in a circle with speed U_A in a counterclockwise direction. At any instant in time, this rotational motion is directed to the right, tangent to the circle in Figure 4.10. If Figure 4.10 represented the Earth, point A would be rotating from west to east. The air parcel is rotating, but it is not physically connected to the turntable. This is akin to the Earth's atmosphere rotating with the planet but not being rigidly bound to it. In response to a push, the parcel begins to move toward the center of the turntable, and let the speed be v measured on the turntable. By analogy to the Earth, this would be a component of motion directed toward the North Pole. The parcel will arrive at the inner circle B in a time t_{AB} defined by the condition $v\, t_{AB} = r_A - r_B$ so that

$$t_{AB} = (r_A - r_B)/v \qquad [4.4.2]$$

As the parcel moves inward toward circle B, it retains its original speed of rotation, U_A. There are now two components to the motion, radially inward and counterclockwise, as viewed by the stationary observer 1.

A point on circle B is rotating with a smaller speed, U_B, than is point A, with speed U_A. Therefore, when the air parcel reaches circle B it will be moving counterclockwise faster than a point that is hooked to circle B. Measured relative to a point hooked to circle B, the parcel will have an eastward speed of $U_A - U_B$. From Eq. 4.4.1, the excess rotational speed is

$$U_A - U_B = \Omega\, (r_A - r_B) \qquad [4.4.3]$$

As viewed by observer 2 connected to point B, the parcel seems to have picked up some counterclockwise speed as it moved from circle A to circle B. Viewed in a reference frame fixed to the rotating turntable, the parcel has accelerated to the right of its initial path. The path of the parcel appears to turn to the right because it has a larger counterclockwise component of speed than its surroundings on circle B. To an observer rotating with the turntable, the parcel appears to follow a curved path that bends toward the right as it moves inward. By analogy to the Earth, the parcel's path turns to the east as viewed from the surface of the planet. As measured from the turntable, the speed of the parcel changes by $U_A - U_B$ in the time t_{AB}, so the acceleration, a_x, is

$$a_x = (U_A - U_B)/t_{AB} \qquad [4.4.4]$$

where, by analogy to the Earth, $+x$ is the coordinate that points east. Use of Eqs. 4.4.2 and 4.4.3 with Eq. 4.4.4 yields a simple result for the acceleration. This is

$$a_x = \Omega v \qquad [4.4.5]$$

The counterclockwise, or eastward, acceleration depends on the product of the rotation rate and the northward speed.

If the air parcel moves outward on the turntable from circle B to circle A, the result is the same. In this case, the parcel at circle B has a relatively small counterclockwise speed as seen by the stationary observer 1. When the parcel moves from circle B to circle A, its counterclockwise speed will be slower than that of the turntable at circle A. Hence, the parcel falls behind a point rotating with circle A. Just as in the earlier case, this corresponds to bending to the right of its initial path as viewed from the rotating turntable. The analogy to the Earth is that an air parcel that starts moving from north to south will develop a component of speed to the west. Regardless of which way the parcel moves, inward or outward, the path turns to the right of its path as viewed by an observer who is rotating with the turntable.

The apparent deflection to the right occurs because wind speeds are measured relative to the surface of a rotating planet, and points on the surface located at different latitudes rotate with different speeds. To explain the movement of air when viewed from this reference frame, it is necessary to invoke a Coriolis force or Coriolis acceleration to explain the motions. If the Earth did not rotate ($\Omega = 0$), there would be no Coriolis force because the reference frame would have an acceleration of zero.

This example assumes that an air parcel began with a component of speed radially inward toward the center of the turntable. In this case, the deflection was to the right. However, the initial component of extra speed could also be tangent to the circle, either eastward or westward using the analogy to the Earth. Let this extra tangential speed of a parcel on circle A be u, initially directed to the east. The direction of u remains fixed as viewed by observer 1 located above the turntable. In addition, the turntable rotates as the parcel moves, and the parcel also has the component of rotational speed U_A around the circle. Figure 4.11 illustrates the situation. At the initial time t_0, u points to the east as viewed from point A, but consider the situation a few hours later at time t_1, after the turntable goes through one quarter of a full rotation. Point A has moved as indicated in Figure 4.11. Observer 1, located above the turntable, clearly sees that the definition of "east" has changed. At time t_0, east pointed to the right, but at time t_1, east points upward in Figure 4.11. This is the case because the coordinate system used by observer 2, who is hooked to the turntable, is rotating. However, the true direction of the extra velocity u is the same as it was in the beginning. Initially this motion was to the east, but after one quarter of a rotation, this same direction in space is defined as south. The speed u has always pointed

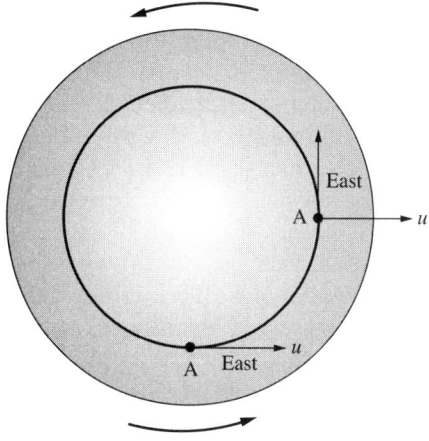

Counterclockwise
Rotation Viewed from Above

FIGURE 4.11   The Coriolis force. When an object moves in a direction initially tangent to the rotating turntable, its path appears to turn to the right. This arises because the directions defined as east, south, etc., are constantly changing.

in the same direction, but the coordinate system used for measuring this direction changes because it is rotating with the turntable. This apparent deflection to the right, from east to south, is another aspect of the Coriolis force. To explain the motion as viewed by observer 2 hooked to the turntable, it is necessary to invoke an acceleration that bends the path of the air parcel to the right of its initial direction of motion. An argument similar to that used previously shows that the acceleration to the right, a_y, depends on the product of the rotation rate of the turntable and the extra speed of the air parcel. The relationship

$$a_y = -\Omega u \qquad [4.4.6]$$

is approximate but still illustrates the nature of the dependence, where the minus sign is necessary because an acceleration to the south is negative. Eqs. 4.4.5 and 4.4.6 show that regardless of which way an object moves on a turntable, it experiences an acceleration to the right of its initial path when viewed from this rotating frame of reference.

Two mechanisms contribute to the Coriolis acceleration. One of these arises from the fact that different points along the radius of a rotating turntable move at different speeds. The other mechanism appears because the coordinate system used to define east, north, and so on is continually changing as the turntable rotates. Equations. 4.4.5 and 4.4.6 arose from simplified examples that view motions as purely radial or purely tangential. In fact, these descriptions fail to capture the full complexity of a rotating coordinate system. Rigorous derivations, as presented in texts on classical mechanics (e.g., Fowles 1962), produce a coefficient of 2Ω in the acceleration in Eqs. 4.4.5 and 4.4.6 instead of Ω alone. However, the essential dependence deduced in the preceding simple cases is correct: the acceleration in one direction depends on the speed in a perpendicular direction.

The Coriolis acceleration may be awkward to describe, but it is a very important influence on winds measured relative to the surface of a rotating planet. It is possible to derive the Coriolis acceleration in a three-dimensional spherical geometry, thereby providing a correct model for air motions on the Earth. The results in Eq. 4.4.5 and 4.4.6, with the additional factor of two noted earlier, are correct in principle, but they must be modified to account for the spherical geometry, in which the Earth's rotation axis is perpendicular to the planet's surface only at the poles. Eqs. 4.4.7 and 4.4.8 are the complete results (e.g., Lorenz, 1967; Holton, 2004):

$$a_y = -2\Omega(\sin \theta)u \qquad [4.4.7]$$

$$a_x = 2\Omega(\sin \theta)v \qquad [4.4.8]$$

where the coordinate x points to the east, y points to the north, and θ is latitude. The speed u is positive when moving west to east, and v is positive moving south to north. In a three-dimensional geometry, the Coriolis acceleration depends on latitude, but Eqs. 4.4.7 and 4.4.8 are qualitatively consistent with the simplified example using the flat turntable. If a parcel is moving northward, so $v > 0$, then the Coriolis acceleration is to the east, and the parcel will develop an eastward component of speed. If a parcel is moving eastward, $u > 0$, it experiences a Coriolis acceleration in the negative y direction, which is to the south. In both cases the path of an air parcel turns to the right of its initial direction. At the equator, $\theta = 0°$, the Coriolis force vanishes, and because of this, tropical wind systems differ in some important respects from those at middle and high latitudes. The fact that the Coriolis force depends on wind speed will be important when it is combined with the pressure gradient force to predict equilibrium winds in the Earth's atmosphere.

The 24-h rotation period of the Earth defines a natural timescale on which the Coriolis force acts. If motions persist for at least a significant fraction of a planetary rotation period, several hours or longer, then the Coriolis acceleration will be significant in determining the nature of these motions. Hence, the Coriolis force is important in studies of weather systems, which persist for days, and global circulation, which varies on the timescale of a season. In contrast, the Coriolis force has little influence on short-lived motions, such as tornadic winds.

4.5   The Centrifugal Force

The centrifugal force is familiar from introductory physics (e.g., Hecht 1996). When a mass m moves in a curved path of radius r, the object experiences a force directed outward, away from the center of rotation. The magnitude of this force is mU^2/r, where U is the object's speed. The centrifugal acceleration is

$$a_{cent} = U^2/r \qquad [4.5.1]$$

As a consequence of the Earth's rotation, a centrifugal force acts on any volume of air that is stationary with respect to the Earth's surface. In this case, the distance of the volume from the axis of rotation is $R \cos \theta$, where θ is latitude and R is the planetary radius, assumed to be very large compared to the height of the volume above the ground. The speed of rotation is $U = \Omega R \cos \theta$, where Ω is the planet's rotation rate defined in Section 4.4. Given these values, Eq. 4.5.1 becomes

$$a_{cent} = \Omega^2 R \cos \theta \qquad [4.5.2]$$

where the direction is perpendicular to the Earth's axis. This total acceleration can be decomposed into a north–south component and a vertical component. The centrifugal acceleration vanishes at the poles, and only the vertical component exists at the equator.

For analysis of the global circulation, one can combine the centrifugal and gravitational accelerations into a single "modified gravitational acceleration." The direction of this resultant force differs by up to 0.1° from that of gravity taken alone (Hess 1959). Furthermore, the direction of the modified gravitational acceleration defines a modified "vertical" coordinate that replaces altitude, whereas new "horizontal" coordinates are perpendicular to this. The desirable outcome is that the centrifugal force associated with the Earth's rotation influences only the vertical balance of forces, using the modified coordinates, and does not appear in the balance that determines horizontal motions. Hess (1959) and Holton (2004) present details of the mathematical transformations that produce this result.

The centrifugal acceleration is a significant influence on horizontal air flows around low- and high-pressure weather systems. Here, U in Eq. 4.5.1 is the horizontal wind speed, measured relative to the ground, as opposed to motion arising from the planet's rotation, and r is the distance from the center of low or high pressure. In this case, the value of r is determined by the dimensions of the weather system rather than by the radius of the planet.

4.6   Friction

The term "friction" refers to any force that opposes motion. The interaction of wind with the Earth's surface provides an important source of atmospheric friction. For example, when wind moves over a body of water, energy is transferred out of kinetic energy possessed by moving air parcels and into the motions of currents and waves. Similarly, when moving air encounters a forest, kinetic energy from the wind is channeled into the motions of tree limbs and, in extreme cases, into the uprooting of trees. It is apparent that these frictional forces are most important in the lowest portion of the atmosphere, a region called the "boundary layer."

Another form of friction, called "viscosity," arises from molecular-scale interactions. Consider, for example, a location where the west-to-east wind varies with altitude. Let the winds at altitude z and $z + dz$ be $u(z)$ and $u(z + dz)$, respectively, where $du/dz > 0$. As described in Section 1.3, molecules in a minute volume of air move about individually, perpetually colliding with their neighbors. When molecules move vertically from z to $z + dz$, where dz is on the order of one mean free path, they have on average a smaller west-to-east component of velocity than their neighbors. Collisions between the "old" and "new" molecules at $z + dz$ act to slow the average bulk speed. Similarly, when molecules from $z + dz$ move downward, collisions act to accelerate the wind at z. Viscosity refers to these interactions that tend to reduce the magnitude of the gradient du/dz. When $du/dz = 0$, the drag on atmospheric motions associated with these molecular-scale interactions vanishes. The role of molecular viscosity can be neglected in many problems involving atmospheric flows, but the above concepts can be extended to describe macroscopic interactions.

Friction with the Earth's surface obviously acts as a drag on atmospheric motions at the ground. But how is this drag communicated throughout the lowest kilometers of the atmosphere? Some atmospheric flows, such as those over irregular terrains, become turbulent, meaning the wind field displays complex variability in space and time. In these cases one can view the erratic motions of macroscopic fluid volumes in a way similar to motions of individual molecules described earlier. The vertical movements of air parcels associated with turbulence exert a drag on the horizontal wind in a way analogous to molecular-scale viscosity. Via this "eddy viscosity," friction with the Earth's surface can influence the vertical structure of horizontal winds throughout the lower kilometers of the atmosphere.

4.7 Several Forces Acting Simultaneously

The pressure gradient force and the Coriolis force acting together explain the general features of the global circulation. To illustrate this, Figure 4.12 depicts a surface of constant altitude in the Northern Hemisphere viewed looking down from above. This horizontal surface could be an altitude of, for example, $z = 5$ km in the middle troposphere. At this altitude, pressure decreases as one moves from low latitudes toward the North Pole. Let $P_1, P_2,$ and P_3 be lines of constant pressure, where $P_1 > P_2 > P_3$. In this situation, the pressure gradient force will cause an air parcel to accelerate from low to high latitudes,

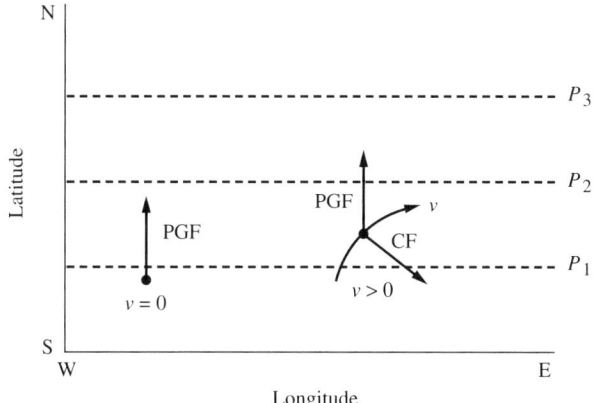

FIGURE 4.12   Horizontal motion of an air parcel experiencing the pressure gradient $(P_1 > P_2 > P_3)$ and Coriolis forces. Straight arrows indicate the direction of the pressure gradient force (PGF) and the Coriolis force (CF). The curved arrow depicts the path of the air parcel.

but as soon as the parcel starts to move, the Coriolis force comes into play. The pressure gradient force always points to the north, and the Coriolis force always points to the right of the instantaneous direction of motion. When the parcel moves to the north, the Coriolis force points to the east. As the parcel moves ever faster, the Coriolis force becomes stronger because it depends on speed. The result is that the parcel bends to the right as observed from a location fixed on the Earth's rotating surface. Figure 4.12 depicts the forces that act on an air parcel at two different times as the speed of the parcel varies. The speed and direction of the air parcel will change until the Coriolis force balances the pressure gradient force as viewed in the rotating frame of reference. This balance occurs when the parcel is moving from west to east, parallel to lines of constant pressure. When this condition of balanced forces is reached, the speed and direction of the air parcel remain constant in time when measured from the Earth's surface. The pressure gradient force acting alone makes air move from high pressure to low pressure, and this seems perfectly logical. Yet, when the pressure gradient and Coriolis forces act together, a very different situation arises. In the balanced state, the wind blows from west to east parallel to lines of constant pressure. This result is not intuitive until one appreciates the nature of the Coriolis force. This balance between the pressure gradient and Coriolis forces is called "geostrophic balance," and the resulting west-to-east wind is called the "geostrophic wind."

When friction is added to this scenario, there is indeed a small wind component directed toward the pole in the middle and upper troposphere. Friction acts to slow the wind speed, so that the Coriolis force alone does not balance the pressure gradient force. In this case, the speed, including both u and v components, adjusts until the sum of all three forces is zero. However, the south-to-north component of flow is very weak compared to the west-to-east wind. The Hadley cell defined previously is real, at least at some latitudes, although the major winds in the global circulation move from west to east parallel to lines of constant pressure. The geostrophic wind plus a contribution from a Hadley circulation provide a reasonably accurate initial description of the global circulation.

The flow of air around low- and high-pressure systems also involves the pressure gradient and Coriolis forces, although the centrifugal force and friction are more important here than in the general circulation. Consider a low-pressure system first. As depicted in Figure 4.13, an air parcel near the ground experiences a pressure gradient force that pushes it toward the center of the low, but as soon as the parcel starts moving, it feels a Coriolis force to the right. In addition, the moving parcel experiences a centrifugal force that always points outward and, in the case of a low, opposes the pressure gradient force. The Coriolis force depends on the speed of the air parcel, which changes until this force balances the pressure gradient and centrifugal forces, so long as friction is negligible. The pressure gradient force always points radially inward, from high pressure to low pressure, whereas the Coriolis force always points to the right of the direction of motion. The three forces can come into balance when the air flows counterclockwise around the center of the low, parallel to the isobars. This counterclockwise motion is called "cyclonic flow." The high-pressure system, also illustrated in Figure 4.13, is the opposite situation. The pressure gradient force here points radially outward, but when an air parcel starts to move in that direction, the Coriolis force bends it to the right. In this case, the centrifugal force serves to reinforce the action of the pressure gradient. The pressure gradient, centrifugal, and Coriolis forces balance each other when the air parcel is moving clockwise parallel to the isobars, a pattern called "anticyclonic flow."

If one considered only the pressure gradient force, the logical conclusion would be that low- and high-pressure systems should self-destruct rapidly.

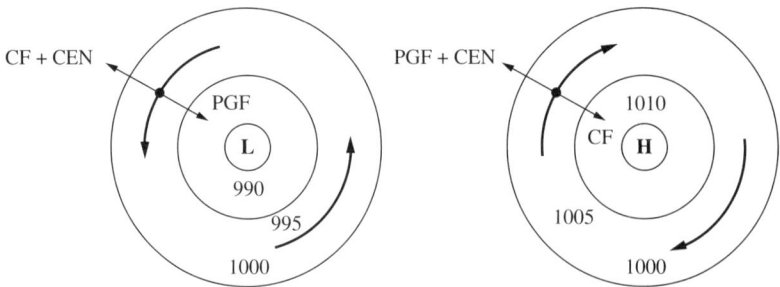

FIGURE 4.13   Horizontal flow of air in low-pressure (left) and high-pressure (right) weather systems. Contours represent surface pressure in millibars. Straight arrows denote directions of the pressure gradient force (PGF), centrifugal force, (CEN) and Coriolis force (CF). Curved arrows indicate wind direction.

For example, air would accelerate into the center of the low, thereby destroying it. However, the Coriolis force leads to circulating motions rather than a purely inward flow. The result is that low-pressure systems can persist for periods of days, although they eventually dissipate.

When friction is present, the described wind patterns change. A low-pressure system illustrates the point. When friction with the ground occurs, the wind speed slows slightly. When air slows, the Coriolis force by itself does not balance the combined pressure gradient and centrifugal forces. Instead, the wind speed adjusts until the sum of all four forces is zero, leading to the wind patterns illustrated in Figure 4.14. The equilibrium state of zero net force is reached when air flows counterclockwise and inward about the center of low pressure. In this case the air has a component of motion across the isobars. Near the ground, parcels of air move in an inward, counterclockwise spiral. Convergence of air toward the center of the low leads to rising motions and promotes cloud formation.

The results in Sections 4.3 and 4.4 provide the basis for a quantitative model of global winds. This model neglects friction, so it applies best at altitudes several kilometers above the ground, where interactions with the local terrain are unimportant. In addition, as stated in Section 4.5, the horizontal component of centrifugal acceleration can be neglected. Eqs. 4.3.6 and 4.3.7 define the northward and eastward accelerations, respectively, of an air parcel subjected only to the pressure gradient force. Equations 4.4.7 and 4.4.8 then specify the corresponding accelerations in response to the Coriolis force. To compute the total acceleration, one need only add the Coriolis accelerations onto the pressure gradient terms. This gives

$$a_y = dv/dt = -(1/\rho) \; dP/dy - 2\Omega(\sin \theta)u \qquad [4.7.1]$$

$$a_x = du/dt = -(1/\rho) \; dP/dx + 2\Omega(\sin \theta)v \qquad [4.7.2]$$

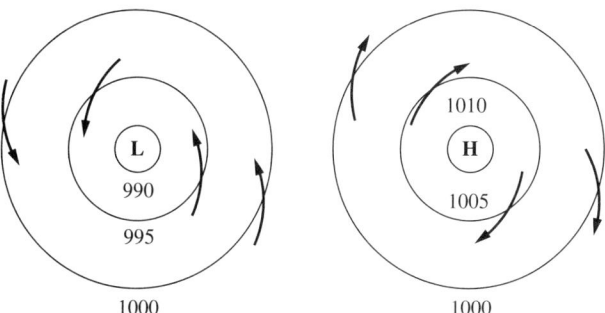

FIGURE 4.14   Horizontal flow of air around low-pressure (left) and high-pressure (right) systems when the pressure gradient force, centrifugal force, Coriolis force, and friction operate. Contours represent pressure in millibars.

Eqs. 4.7.1 and 4.7.2 are the equations of motion on a rotating planet. Apart from friction, they provide a quantitative model of the general circulation when the pressure gradients are known.

The concept of motion under balanced forces requires that the velocity of an air parcel change until the net force, or the acceleration, is zero. With accelerations set to zero, Eqs. 4.7.1 and 4.7.2 produce the south-to-north (v) and west-to-east (u) wind speeds in response to known west-to-east and south-to-north pressure gradients, respectively. The west-to-east wind is

$$u = - [\, 2\Omega(\sin\,\theta)\,\rho]^{-1}\,dP/dy \qquad\qquad [4.7.3]$$

and the south-to-north wind is

$$v = [\, 2\Omega(\sin\,\theta)\,\rho]^{-1}\,dP/dx \qquad\qquad [4.7.4]$$

Heating of the Earth's surface by sunlight creates and maintains a pressure gradient between the equator and the poles. Pressure at a constant altitude decreases as latitude increases in the middle to upper troposphere, so that $dP/dy < 0$. Eq. 4.7.3 then requires $u > 0$, corresponding to flow from west to east. This is the "geostrophic wind."

Eq. 4.7.3, if considered alone, leads to a problem in describing the atmospheric energy balance. Chapter 2 showed that there is a net heating of low latitudes by sunlight, and there is a net radiative cooling of higher latitudes, where the loss of terrestrial energy exceeds absorption of solar energy. The temperature distribution associated with this latitudinal imbalance between radiative heating and cooling creates the pressure gradient $dP/dy < 0$. This pressure gradient, if acting alone, attempts to drive a circulation to transport heat from equator to pole to compensate for the radiative imbalance, as would be done by the Hadley circulation. However, the Coriolis force introduces a major complication. As a result of the Coriolis force, a north-to-south pressure gradient, $dP/dy < 0$, leads to a wind that blows from west to east, and a wind that blows from west to east obviously does not transport heat from equator to pole. Yet, the atmosphere needs to accomplish this heat transport to maintain an energy balance at each latitude. Otherwise, based on radiation alone, the equatorial regions keep getting hotter, and the high latitudes keep getting colder.

How does the atmosphere accomplish the required transport of heat? Eq. 4.7.4 contains the answer. A wind from south to north will arise from a west-to-east pressure gradient. Based on latitudinal temperature differences, it is easy to see why there should be a pressure gradient $dP/dy < 0$ between the equator and the poles. However, a pressure gradient in the west-to-east direction is another matter. To illustrate this, Figure 4.15 depicts a section of the Earth viewed from above. Why should the pressure at point C in Figure 4.15 be different from the pressure at point D when both points lie at the

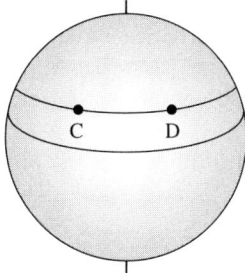

FIGURE 4.15   A section of atmosphere at fixed altitude, where points C and D are at the same latitude. The pressure at C can differ from that at D because of the presence of low-pressure and high-pressure weather systems.

same latitude? The answer is that low- and high-pressure weather systems can provide pressure differences that lead to a nonzero value of dP/dx at any given point.

Consider a fixed latitude such as 40° north and an altitude of, for example, 3 km. Imagine measurements of pressure at this altitude plotted as a function of longitude all around the globe at constant latitude. These measurements would show a wavy pattern, as illustrated in Figure 4.16. Regions of higher-than-average pressure alternate with regions of lower pressure. This wave structure is related to the presence of high- and low-pressure weather systems. The pressure gradient is positive from one trough to the next peak. According to Eq. 4.7.4, this drives a wind going south to north.

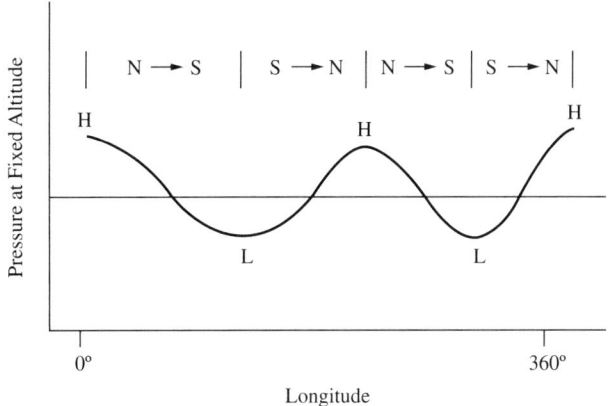

FIGURE 4.16   Atmospheric pressure at a fixed altitude and latitude as a function of longitude. The alternating pattern of highs and lows leads to winds to the north and to the south at different longitudes.

Then there is a region where the pressure gradient is negative, from a peak to a trough, and this is a region where winds move north to south. The observed pressure distribution combined with the constraint of balanced forces requires that the north–south winds alternate in sign as one moves around the globe.

The geographic pattern of high and low pressure changes from day to day as weather systems move and evolve. As a result, the north–south winds vary in an erratic way, but the net result of these motions is the transport of heat from equator to pole. The Hadley cell provides a simple model of north–south motions in the troposphere, and a Hadley-type circulation does indeed exist at low latitudes, within about 30° of the equator, where the Coriolis force is relatively weak. However, at middle and high latitudes, the south-to-north transport of heat is accomplished primarily by low- and high-pressure weather systems. The collective action of these weather systems is critical to maintaining the latitudinal energy balance of the atmosphere. Palmen and Newton (1969) give a detailed account of these circulations, and Lutgens and Tarbuck (2004) provide an introductory summary of the range of wind systems encountered in the atmosphere.

4.8 Observed Winds

The previous sections demonstrate that vertical motions and north–south, or "meridional motions," can vary spatially and temporally in complex ways, where continuity of mass and the need to transport heat from low to high latitudes provide the ultimate constraints. An interpretation of the west–east, or "zonal," flow is more straightforward. These motions arise from the seasonally varying pressure gradient in latitude, and the relatively large wind speeds are readily measured and provide a sound basis for statistical summaries.

Figure 4.17 depicts the zonal wind u averaged over longitude for the Northern Hemisphere in winter (top panel), defined as December–January–February, and summer (bottom panel), encompassing June–July–August. The values are as given by Lorenz (1967). Although newer data are available, the general features shown here have been well documented for more than half a century. Note that Figure 4.17 uses pressure as the vertical coordinate, thereby highlighting the troposphere, which contains approximately 80% of the atmosphere's mass. A positive value of u in m s^{-1} refers to flow from west to east and via Eq. 4.7.3 corresponds to a pressure at fixed altitude that declines from low latitudes toward the pole, $dP/dy < 0$.

Figure 4.17 shows that winds moving from west to east, or "westerlies," exist at middle to high latitudes of both hemispheres, consistent with prevailing pressure and temperature gradients. Note from Eq. 1.9.12 that horizontal gradients in pressure can arise from spatial differences in temperature.

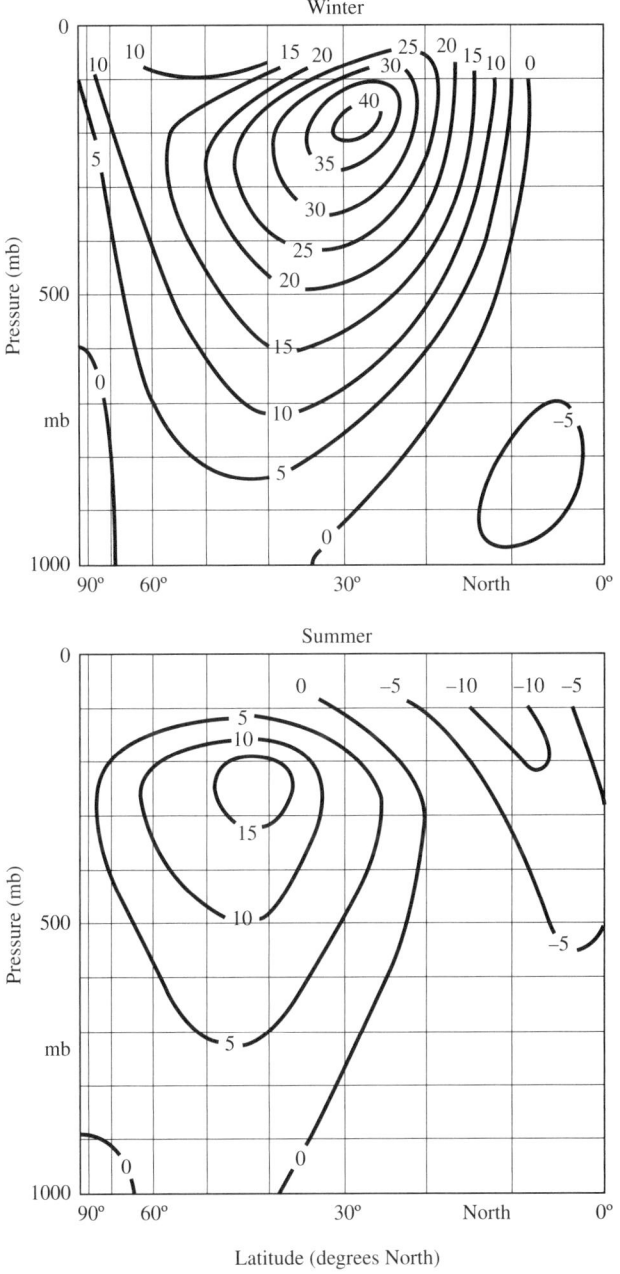

FIGURE 4.17 Average zonal wind u as a function of latitude and atmospheric pressure. Top panel: winter, average of December–January–February. Bottom panel: summer, average of June–July–August. Contours are in m s^{-1} based on data by Lorenz (1967).

The westerlies reach a maximum in the upper troposphere, where the core of strongest winds is the "jet stream." During summer the jet stream maximum lies at latitudes from about 35° north to 50° north. The wintertime maximum shifts to lower latitudes, 25° to 35° north, and has a greater magnitude than in summer. This reflects the larger latitudinal temperature gradient that exists in winter.

The absence of a substantial seasonal variation in solar heating in the tropics leads to relatively small gradients and light winds in this region. However, the weak east-to-west winds, or "easterlies," that exist in both summer and winter and that extend to approximately 35° north at the surface require explanation. Intense heating by sunlight drives vertical convection in the tropics, with the rising air parcels reaching high altitudes in the troposphere. The region of intense convection migrates seasonally, the subsolar point being near 20° north and 20° south for the top and bottom panels, respectively, in Figure 4.17. Near the ground, air flows toward the tropics from approximately 30° to 35° north to compensate for the upward flow. The accompanying Coriolis deflection to the right produces a wind component from east to west at these latitudes.

This interpretation is "diagnostic" in the sense that, given known horizontal distributions of pressure, the geostrophic wind equation provides reasonable estimates of seasonally averaged zonal winds. The pressure gradients are themselves the result of temperature gradients produced by latitude-dependent heating and cooling rates, but, in fact, the global circulation is more complex than this picture suggests. This is because winds that develop in response to pressure and temperature gradients act to alter the gradients that created them. A first-principles "prognostic" estimate of atmospheric winds requires the simultaneous and self-consistent calculation of all of these interdependent quantities as functions of location and time.

If one assumes hydrostatic balance (Eq. 1.9.12), the relationship between pressure and temperature is straightforward. However, a complete theoretical treatment of temperature requires application of the First Law of Thermodynamics (Section 3.7) to an atmosphere that is warmed by both radiation and latent heat release and where vertical motions lead to adiabatic heating and cooling. Chapters 2 and 3 considered radiation, adiabatic motions, and latent heat release separately, but the true distribution of atmospheric temperature clearly reflects their simultaneous action. In addition, horizontal winds can cause the temperature at a fixed point to change by transporting air parcels with different thermal characteristics to that location from elsewhere. Further complications arise from the interaction of winds with the varying nature of the Earth's surface. Mountain ranges act as obstacles to the horizontal motion of air and thereby create disturbances that appear as longitude-dependent wave patterns imposed on the zonal flow. Finally, the flow of air over oceans is accompanied by the dissipation of wind energy to drive water currents, and evaporation provides the vapor that will eventually be a source of latent heat to the atmosphere. The result is an interconnected air–ocean–solid Earth system that supports the huge variety of motions observed in the atmosphere.

4.9 Exercises

The following information is relevant to all of the exercises for this chapter. The radius of the Earth is $R = 6371$ km, and the rotation rate of the planet on its axis is $\Omega = 7.27 \times 10^{-5}$ s^{-1}. This corresponds to one full revolution in 24 h. Tables 4.2 and 4.3 give the pressure (P), temperature (T), and mass density (ρ) of the atmosphere at an altitude of $z = 5$ km at various latitudes for the months of January and July (*U.S. Standard Atmosphere Supplements, 1966*). A metric unit of pressure is dynes cm^{-2}, where 1000 dynes cm$^{-2} =$ 1 mb.

1. Eq. 4.5.2 defines the centrifugal acceleration experienced by an object that is stationary with respect to the Earth's surface at latitude θ. This total acceleration is directed perpendicular to the Earth's rotation axis and can be decomposed into components in the north–south and vertical directions.

    a. The upward component of the centrifugal acceleration is $\Omega^2 R$ $(\cos θ)^2$. This upward force opposes the downward force due to gravity. One can define the total vertical acceleration as $g' = g - \Omega^2 R$ $(\cos θ)^2$, where $g = 9.81$ m s^{-2}. Consider latitude 40° north. What is the percent difference between g and g'?

    b. The north-to-south component of the centrifugal acceleration is $\Omega^2 R$ $(\cos θ)(\sin θ)$, and this is perpendicular to the vertical acceleration g' computed in (a). Define a "modified gravitational acceleration" as the vector sum, accounting for both magnitude and direction, of g' and $\Omega^2 R$ $(\cos θ)(\sin θ)$. What is the value of the "modified gravitational acceleration" at latitude 40° north, and what is the angle between this acceleration and that due to gravity alone?

TABLE 4.2 Model Atmosphere at an Altitude of 5 km for January (Northern Hemisphere)

Latitude (°N)	Pressure (mb)	Temperature (Kelvin)	Density (g cm^{-3})
15	559	271	7.20×10^{-4}
30	552	262	7.34×10^{-4}
45	531	250	7.41×10^{-4}
60	516	241	7.46×10^{-4}

TABLE 4.3 Model Atmosphere at an Altitude of 5 km for July (Northern Hemisphere)

Latitude (°N)	Pressure (mb)	Temperature (Kelvin)	Density (g cm^{-3})
15	559	271	7.20×10^{-4}
30	559	272	7.16×10^{-4}
45	554	267	7.21×10^{-4}
60	541	260	7.24×10^{-4}

2. Estimate the west–east component of the geostrophic wind, u, in m s^{-1} for the three latitude bands 15°–30°, 30°–45°, and 45°–60° north for both January and July. To do this, assume that pressure gradients computed from values in Tables 4.2 and 4.3 apply to the midpoint of each latitude band. For example, the gradient computed using pressures at 30° north and 45° north can be associated with latitude 37.5° and taken to represent an average for the 15°-wide band. Be sure to state whether the wind is moving from west to east or from east to west.

3. If the results in Exercise 2 are correct, the winds in winter (January) are stronger than those in summer (July). This arises because of different pressure gradients during the two months. Suggest a physical explanation for this seasonal difference in pressure gradients.

4. The wind calculation in Exercise 2 involved differences between atmospheric pressures at two latitudes. These differences were small compared to the absolute values of the pressures themselves. Suppose there was a random error of plus or minus 2% on the measured pressures in Tables 4.2 and 4.3. This means that the true atmospheric pressure at a point may be anywhere from 2% smaller to 2% larger than the value listed, and the exact value of the error can differ from one latitude to another. Discuss qualitatively the effect such a random error in pressure would have on the geostrophic wind calculations. This exercise illustrates why wind calculations pose a large practical problem unless methods exist for obtaining highly accurate pressures.

4.10 References

Fowles, G. R. 1962. *Analytical Mechanics*. New York: Holt, Rinehart and Winston.

Goody, R. M., and J. C. G. Walker. 1972. *Atmospheres*. Englewood Cliffs, N.J.: Prentice–Hall.

Hecht, E. 1996. *Physics: Calculus*. Pacific Grove, Calif.: Brooks Cole Publishing Co.

Hess, S. L. 1959. *Introduction to Theoretical Meteorology.* New York: Holt, Rinehart and Winston.

Holton, J. R. 2004. *An Introduction to Dynamic Meteorology.* New York: Academic Press.

Lorenz, E. N. 1967. *The Nature and Theory of the General Circulation of the Atmosphere.* Geneva: World Meteorological Organization.

Lutgens, F. K., and E. J. Tarbuck. 2004. *The Atmosphere—An Introduction to Meteorology.* Upper Saddle River, N.J.: Pearson Prentice–Hall.

Palmen, E., and C. W. Newton. 1969. *Atmospheric Circulation Systems: Their Structure and Physical Interpretation.* New York: Academic Press.

U.S. Standard Atmosphere Supplements, 1966. Washington, D.C.: U.S. Government Printing Office, 1966.

Chapter 5

Chemical Processes and Atmospheric Ozone

INTRODUCTION

Atmospheric chemistry begins when molecules dissociate after absorbing solar energy, especially in the ultraviolet portion of the spectrum. The constituents so created participate in a series of chemical reactions that determine the trace gas composition of the atmosphere, where the abundance of each gas represents a balance between production and loss. The formation of the stratospheric ozone layer is a classic problem in atmospheric chemistry. The dissociation of molecular oxygen is the initial step that ultimately leads to the creation of ozone, whereas loss occurs in catalytic cycles involving chlorine and oxides of nitrogen. The accidental leakage of manmade chlorofluorocarbons from refrigeration equipment during the late twentieth century provided a mechanism for chlorine to reach the stratosphere and, therefore, a means for human activity to influence the ozone abundance. The dramatic decline in springtime ozone amounts over Antarctica is an extreme case in which anthropogenic chlorine had an unexpectedly large impact on the lower stratosphere.

The release of hydrocarbons, carbon monoxide, and nitrogen oxides at the ground is an unintentional byproduct of the combustion of fossil fuels. These primary pollutants participate in a series of chemical reactions that can produce elevated summertime ozone amounts at the ground in urban areas. Technological advances have reduced the magnitude of these undesirable emissions, and air quality in many cities has improved in recent decades.

5.1 Atmospheric Chemistry: An Overview

Absorption of solar radiation provides the energy source to drive a wide range of chemical processes over altitudes ranging from the thermosphere to the ground. Sunlight in the ultraviolet portion of the spectrum is central to atmospheric chemistry. This reflects the fact that the energy per photon must be sufficiently large to break typical atmospheric molecular bonds. At altitudes greater than 90 km, ultraviolet radiation at wavelength shorter than 100 nm leads to production of positive ions and free electrons that influence radio communications. Longer wavelength ultraviolet light, extending from

100 nm to roughly 300 nm, initiates the chemistry of the mesosphere and stratosphere, the most important consequence being the Earth's ozone layer. Finally, radiation at wavelengths near and longer than 300 nm is absorbed either in the troposphere or at the ground. Various chemical processes in the troposphere are associated with urban air pollution, which begins with products of combustion such as hydrocarbons and carbon monoxide that lead to creation of ozone and other toxic species near the ground in affected regions.

Atmospheric chemistry is devoted to understanding mechanisms that determine the abundances of trace gases that reside at all altitudes. Theoretical investigations require, first, identifying all chemical reactions that are important to the problem at hand and, second, converting this set of reactions into a quantitative model to compute the abundances of various trace gases. To proceed with this second task, it is necessary to define production rates, loss rates, and the rate coefficients for the types of chemical processes that occur in the atmosphere. Three categories of gas-phase reactions require consideration, and some problems involve additional interactions between the gas and liquid or solid phases. The gas-phase processes are "one-body," "two-body," and "three-body" reactions. Here, the symbols A, B, C, D, and M may refer to either atoms or molecules, and AB represents a molecule composed of two or more atoms.

The most important atmospheric one-body reactions are "photodissociations." Here a molecule, denoted by AB, absorbs one photon and splits apart into its components:

$$AB + h\nu \rightarrow A + B \qquad [R\text{-}1]$$

where $h\nu$ denotes a photon with sufficient energy to break the bond in AB. In this chapter, reactions will be labeled as R-1, R-2, etc., to distinguish them from equations. The loss rate of AB, with the dimensions of molecules per unit volume per unit time, is numerically equal to the production rate of A and the production rate of B but of a different sign. These quantities are

$$-d[AB]/dt = d[A]/dt = d[B]/dt = J[AB] \qquad [5.1.1]$$

where t is time and J, in s^{-1}, is the rate coefficient for the one-body process. In general, the number density of a gas in cm^{-3} is indicated by its chemical symbol in square brackets, such as [AB], [A], and [B]. The value of J, the "dissociation rate," depends on the flux of sunlight at wavelengths capable of breaking the bond in AB as well as on the energy level structure of the molecule. Details appear in Section 5.2.

Next consider a general two-body reaction of the form

$$A + B \rightarrow C + D \qquad [R\text{-}2]$$

where A and B are the "reactants" and C and D are the "products." The two-body rate coefficient k is defined by

$$-d[A]/dt = -d[B]/dt = d[C]/dt = d[D]/dt = k[A][B] \qquad [5.1.2]$$

The production and loss rates are proportional to the two-body rate coefficient and the number densities of all of the reactants. Typical units of a two-body rate coefficient are $cm^3\,s^{-1}$. Note that production and loss rates depend only on the number densities of the species that are colliding and reacting with each other, that is, on the species on the left-hand side of a reaction.

Finally, a three-body reaction has the general form

$$A+B+M \rightarrow AB+M \qquad [R\text{-}3]$$

Reaction R-3 is a shorthand notation for two different two-body processes that occur in rapid succession. These are

$$A+B \rightarrow AB^* \qquad [R\text{-}4]$$

and

$$AB^*+M \rightarrow AB+M \qquad [R\text{-}5]$$

The * symbol in R-4 and R-5 denotes an AB molecule with a large amount of internal energy in the form of vibrations of its component atoms. If the molecule does not rid itself of this excess energy very rapidly, it will break apart into separate A and B to yield no net effect. An effective way to dispose of this energy is for AB* to transfer it to another molecule in a collision, as depicted in R-5. The identity of the atom or molecule M is unimportant; its only role is to carry away the excess energy to produce a stable AB molecule. In the Earth's atmosphere, M is most likely to be N_2 or O_2. It is convenient to depict the combined effects of R-4 and R-5 as the single three-body process R-3. The rate coefficient for R-5, typically in $cm^6\,s^{-1}$, is defined by

$$-d[A]/dt = -d[B]/dt = d[AB]/dt = k[M][A][B] \qquad [5.1.3]$$

Note that the third body, M, is neither created nor destroyed.

In principle, numerical values of rate coefficients for two-body and three-body reactions can be computed by considering molecular-scale interactions that occur when reactants approach each other. The simplest scenario is one in which each collision between one A and one B leads to a reaction. When this is the case, there is a simple relationship between the collision frequency of the reactants and the rate coefficient. In practice, the situation is far more complex. Repulsive forces can arise as the outermost electron orbits of the reactants begin to overlap, and these can prevent a reaction in all but a small fraction of the encounters. Owing to these complications, rate coefficients for two-body and three-body reactions typically are deduced empirically by mixing the appropriate gases together and recording changes in number densities over time. Numerical values of k result from application of Eqs. 5.1.2 or 5.1.3.

Rate coefficients for different reactions vary over a wide range, depending on structural details of the colliding atoms or molecules. If a reaction occurred on every collision between the species involved, the rate coefficient for a two-body process would be approximately 2×10^{-10} cm^3 s^{-1}. In practice, most rate coefficients are substantially smaller than this upper limit, and many have a strong inverse dependence on temperature because of the repulsive forces identified here. A typical three-body rate coefficient is on the order of 10^{-30} cm^6 s^{-1} or perhaps several orders of magnitude less.

5.2 Absorption of Solar Radiation: The Photodissociation Rate

The calculation of dissociation rates, J in Eq. 5.1.1, is central to models of atmospheric chemistry. To see what is involved here, consider the photodissociation R-1. One can view the molecule AB as analogous to two golf balls, representing atoms A and B, held together by a spring, which accounts for the bond. For molecules that consist of three atoms, one of A or B will itself be a molecule. For example, in the case of O_3, A can refer to O, while B will be O_2. The components A and B vibrate back and forth about an equilibrium separation as the spring alternately stretches and compresses, and discrete energy states of the bound molecule correspond to specific allowed amplitudes of these vibrations. However, if sufficient energy is added to the molecule, the spring can break, yielding the products. If the molecule absorbs one photon with the required energy, a photodissociation occurs.

The minimum energy needed to break a molecular bond is called the "dissociation energy," labeled by D, where the "electron volt" (ev) is a convenient unit. Typical dissociation energies for atmospheric molecules are in the range 1 to 10 ev, where 1 ev $= 1.602 \times 10^{-19}$ J.

Equation 2.1.1 can define a "dissociation wavelength" λ_D as

$$D = hc/\lambda_D \qquad [5.2.1]$$

When D is expressed in ev and λ_D is in nm, the numerical relationship becomes

$$D(\text{ev}) = 1240/\lambda_D(\text{nm}) \qquad [5.2.2]$$

A necessary, but not sufficient, condition for dissociation is that the energy of the photon, E, satisfy $E > D$, or equivalently that the wavelength satisfy $\lambda < \lambda_D$. Table 5.1 presents dissociation energies and wavelengths for the

TABLE 5.1 Dissociation Energies (D) and Dissociation Wavelengths (λ_D) for Molecular Oxygen and Ozone

Dissociation	D (ev)	λ_D (nm)
$O_2 + h\nu \rightarrow O + O$	5.12	242
$O_3 + h\nu \rightarrow O_2 + O$	1.05	1181

photodissociations of molecular oxygen and ozone, two processes that are central to the chemistry of the Earth's atmosphere.

As an example, consider the photodissociation of O_2:

$$O_2 + h\nu \rightarrow O + O \qquad \text{[R-6]}$$

which is energetically possible for ultraviolet wavelengths shorter than 242 nm. The sun provides a certain number of photons per second in this wavelength range at the top of the atmosphere. As these photons move downward, they encounter O_2 molecules in increasing number densities as altitude decreases. In passing through a layer of a specific vertical thickness, some fraction of the photons incident at the top of this layer are absorbed and lead to dissociation. At wavelengths greater than $\lambda_D = 242$ nm, absorption by O_2 does not occur, whereas at $\lambda < 242$ nm, the fraction of incident photons absorbed varies with wavelength. In treating the absorption, it is necessary to account for this wavelength dependence.

As photons are absorbed, the energy flux contained in the solar beam shrinks. The magnitude of the decrease depends on the abundances of the absorbing molecules and on the strength of the absorption per molecule as a function of wavelength, a concept that is defined rigorously later. Finally, the dissociation rate of a molecule at a given altitude depends on an integral over all wavelengths that correspond to photons capable of breaking the bond. The decrease in solar energy flux with decreasing altitude leads to similar behavior in the dissociation rate. The following paragraphs put this reasoning on a quantitative basis with the goal of deriving an explicit expression for the dissociation rate J in Eq. 5.1.1.

Figure 5.1 depicts a beam of sunlight propagating downward through the atmosphere at an angle θ from the vertical. The thin layer between altitudes z-dz and z contains several different types of molecules that absorb solar radiation at wavelengths between λ and $\lambda + d\lambda$. Let the number densities of these absorbers be $n_A(i,z)$, $i = 1, 2, \ldots, N$. In practice, molecular oxygen and ozone provide the only important attenuation at altitudes less than 90 km, although for later use it is essential that the following derivation include all absorbers. Let $\mathbb{E}(\lambda,\theta,z)\,d\lambda$ represent the solar energy per unit time in the wavelength range λ to $\lambda + d\lambda$ that crosses a unit area oriented perpendicular to the energy flow. Typical units for $\mathbb{E}(\lambda,\theta,z)$, the energy flux per unit

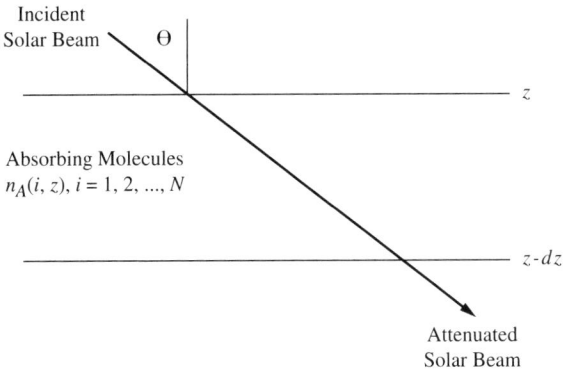

FIGURE 5.1   Attenuation of a beam of sunlight as it passes downward through an atmospheric layer extending from altitude z to z-dz. The layer contains several gases that absorb solar energy at the wavelength of interest.

wavelength, are W m^{-2} nm^{-1}, where wavelength is expressed in nm. As energy moves downward from z to z-dz it is attenuated, where absorption leads to dissociation. The change in radiant energy flux per unit wavelength across the layer, $d\mathbb{E}(\lambda,\theta,z) = \mathbb{E}(\lambda,\theta,z\text{-}dz) - \mathbb{E}(\lambda,\theta,z) < 0$, is proportional to the incident energy flux, the slant-path distance taken by the beam through the layer, $dz/(\cos\theta)$, and the weighted sum of the number densities of the absorbing gases. The relationship is

$$d\mathbb{E}(\lambda,\theta,z) = -\left\{ \sum_{i=1}^{N} \sigma_A(i,\lambda)n_A(i,z) \right\} \mathbb{E}(\lambda,\theta,z)\, dz/(\cos\theta) \qquad [5.2.3]$$

where the summation extends over all absorbing gases ($i = 1, 2, \ldots, N$) and $\sigma_A(i,\lambda)$ is the "absorption cross section" of gas i at wavelength λ.

The absorption cross section, whose dimension is area, specifies the strength and wavelength dependence of the attenuation provided by a molecule. When λ is longer than the dissociation wavelength λ_D, the cross section is zero, whereas when $\lambda < \lambda_D$, the cross section has a magnitude and wavelength dependence that depend on details of molecular structure. In principle, absorption cross sections can be computed from a comprehensive knowledge of molecular structure. In practice, values of cross sections used in atmospheric models are measured in the laboratory by directing light of various wavelengths through a vessel containing a known quantity of the molecule of interest. Values of σ_A are deduced by measuring the attenuation as a function of wavelength.

The solution of Eq. 5.2.3 for the energy flux per unit wavelength at altitude z is

$$\mathbb{E}(\lambda,\theta,z) = \mathbb{E}_0(\lambda) \exp\left\{ -\sum_{i=1}^{N} \sigma_A(i,\lambda) \int_0^z n_A(i,z')\, dz'/(\cos\theta) \right\} \qquad [5.2.4]$$

where $\mathbb{E}_0(\lambda)$ is the incident solar flux per unit wavelength at the top of the atmosphere, and the unit area across which both $\mathbb{E}(\lambda,\theta,z)$ and $\mathbb{E}_0(\lambda)$ flow is oriented perpendicular to the direction of propagation. Equation 5.2.4 is a form of Beer's Law applicable to the atmosphere. The quantity $\mathbb{E}_0(\lambda)$ is related to the solar constant introduced in Section 2.3 by $S_E = \int_0^\infty d\lambda\, \mathbb{E}_0(\lambda)$. For future use, the flux of energy per unit wavelength crossing a horizontal area, based on the discussion in Section 2.9, is

$$\mathbb{E}_H(\lambda,\theta,z) = (\cos\theta)\, \mathbb{E}(\lambda,\theta,z) \qquad [5.2.5]$$

An expression for the photodissociation rate, J in Eq. 5.1.1, follows from Beer's Law combined with concepts from the particle model of light introduced in Chapter 2. The central idea is that one molecule can dissociate when it absorbs one photon whose energy exceeds the dissociation energy or, equivalently, whose wavelength is less than λ_D. Consider a unit volume of vertical thickness dz in the horizontal layer depicted in Fig. 5.1. The energy in the wavelength range λ to $\lambda + d\lambda$ absorbed per unit volume in this layer is $[d\mathbb{E}_H(\lambda,\theta,z)/dz]\, d\lambda$. The number of photons absorbed per unit volume per second in this wavelength range is $(hc/\lambda)^{-1}\,[d\mathbb{E}_H(\lambda,\theta,z)/dz]\, d\lambda$, where the factor hc/λ, from Eq. 2.1.1, is the energy of one photon.

Let AB identify a molecule that absorbs and dissociates in the wavelength range of interest. Given this, AB is one term, say, the jth, in the summation of Eq. 5.2.4. The number of dissociations of AB per unit volume per second in the wavelength interval λ to $\lambda + d\lambda$ is, by analogy to the reasoning presented, $dL = (hc/\lambda)^{-1}\,[d\mathbb{E}_H(\lambda,\theta,z)/dz]_j\, d\lambda$, where the subscript j denotes the rate of change of $\mathbb{E}_H$ with altitude due to absorption by AB alone. The total number of dissociations of AB occurring per volume per second is an integral over all wavelengths:

$$L = \int_0^\infty (hc/\lambda)^{-1}\,[d\mathbb{E}_H(\lambda,\theta,z)/dz]_j\, d\lambda \qquad [5.2.6]$$

Use of Eqs. 5.2.4 and 5.2.5 yields

$$L = \left[\int_0^\infty (hc/\lambda)^{-1}\,\sigma_A(j,\lambda)\,\mathbb{E}(\lambda,\theta,z)\, d\lambda\right] n_A(j,z) \qquad [5.2.7]$$

Since $n_A(j,z)$ is identical to the number density [AB] in Eq. 5.1.1, the dissociation rate of molecule j is

$$J_j = \int_0^\infty (hc/\lambda)^{-1}\,\sigma_A(j,\lambda)\,\mathbb{E}(\lambda,\theta,z)\, d\lambda \qquad [5.2.8]$$

For use in computing dissociation rates it is common to express the solar flux in the units photons $m^{-2}\, s^{-1}\, nm^{-1}$ via $F(\lambda,\theta,z) = (hc/\lambda)^{-1}\,\mathbb{E}(\lambda,\theta,z)$, so that

$$J_j = \int_0^\infty \sigma_A(j,\lambda)\,F(\lambda,\theta,z)\, d\lambda \qquad [5.2.9]$$

with σ_A expressed in m^2 and λ in nm, J_j has the units s^{-1}.

In practice the upper limit of the integration over wavelength is λ_D since $\sigma_A(j,\lambda) = 0$ when $\lambda > \lambda_D$. For calculations in the Earth's mesosphere and below, the lower limit on the integration can be set to 100 nm since radiation at shorter wavelengths is consumed at thermospheric altitudes. In principle, the calculation of dissociation rates using Eqs. 5.2.4 and 5.2.8 or 5.2.9 requires knowledge of the vertical profile and absorption cross section of each molecule that absorbs over the wavelength range of interest. However, for applications to altitudes below 90 km, it is sufficient to include only the attenuation provided by molecular oxygen and ozone.

5.3 Atmospheric Ozone

The existence of ozone in the Earth's atmosphere provides an important case study of an atmospheric chemical system. The ozone molecule, O_3, appears in atmospheric chemistry in two different ways, both of which couple to human health. In one case, a decrease in the abundance of ozone in the stratosphere could, over a period of years, lead to negative biological effects at the Earth's surface. This arises because a decline in the total number of ozone molecules in a column of the atmosphere would be accompanied by an increase in the flux of ultraviolet sunlight reaching the ground, and this radiation can harm living tissues. On the other hand, industrialized areas occasionally experience periods when relatively large amounts of ozone exist at the ground. These increases in ground-level ozone abundances can also have adverse impact on human health since excessive amounts of ozone are known to aggravate existing respiratory problems. Ozone is "good" if it resides far above the ground and acts as a shield against solar ultraviolet radiation, but ozone is "bad" when excessive quantities come into contact with the respiratory system.

Ozone in the stratosphere has received considerable attention because of an unexpected decline in the abundance of this trace gas that occurs over Antarctica each austral spring. The changes, however, are not confined to Antarctica. It is firmly established that stratospheric ozone amounts over middle latitudes declined for two to three decades after 1970. For example, over the latitude band from $26°$ north to $64°$ north and for the years 1970 to 1991, the downward trend derived from ground-based measurements varies from about -1.0% per decade to -2.7% per decade depending on season (World Meteorological Organization 1991), where these percentages refer to the total number of ozone molecules in a column of the atmosphere.

Ozone is an excellent example of a trace gas that has a significant influence on the habitability of the planet. Only about 1 in every 2 million molecules in the Earth's atmosphere is ozone. Yet, this small amount is sufficient to filter out much of the sun's ultraviolet radiation at wavelengths shorter than 320 nm. The major questions concerning stratospheric ozone to be addressed in this chapter are, How is it created and destroyed? and, How have human activities led to a change in the amount of ozone in the stratosphere? The chapter ends with similar questions concerning ozone at the ground in urban areas.

Observations of the radiant energy emitted by very hot objects in the laboratory provide reasonable estimates of the spectrum of sunlight incident at the top of the Earth's atmosphere. As described in Chapter 2, most of the sun's energy lies in the visible and near-infrared regions; however, about 8% of the total energy flux is in the ultraviolet. Yet, when scientists measured the spectrum of solar radiation reaching the ground, they found that the flux in the ultraviolet portion was not as expected. There was much less ultraviolet solar energy at the Earth's surface than expected from the emission of hot lamps. Figure 5.2 illustrates the spectrum of ultraviolet solar flux at the ground. At wavelengths shorter than 330 nm, there is a steep drop in energy, and at wavelengths less than about 300 nm, essentially no solar energy reaches the ground.

The missing solar energy posed a problem until 1881, when William Hartley proposed the correct explanation. Hartley subjected a container of air in the laboratory to electrical sparks and found that ozone was created in the process. He then directed ultraviolet light from a lamp through the container of ozone. He knew the spectrum of light emitted by the lamp, and he could measure the spectrum transmitted through the gas. These observations revealed that ozone absorbed the very same wavelengths that were ab-

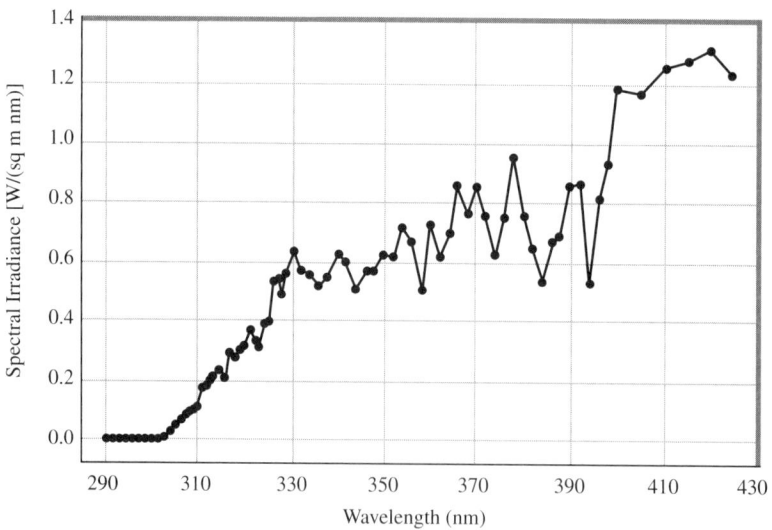

FIGURE 5.2   A typical spectrum of solar ultraviolet energy flux, also called "spectral irradiance," as measured at the ground. The sharp drop in energy flux at wavelengths shorter than 330 nm arises from absorption by ozone located primarily in the stratosphere. [Based on Frederick (2003).]

sent from sunlight at the Earth's surface. Based on these measurements, Hartley proposed that ozone existed in the Earth's atmosphere and that it absorbed the ultraviolet sunlight that did not appear at the ground. Today the rapid drop in solar ultraviolet flux at the ground is called the "ozone cut-off." Craig (1965) has given a detailed account of the early history of studies related to stratospheric ozone.

5.4   Chemistry in a Pure-Oxygen Atmosphere

In 1930, Sydney Chapman proposed the first rigorous theory of the ozone layer's formation (Chapman 1930), and this work marked the beginning of modern atmospheric chemistry. The major realization is that molecular oxygen, O_2, breaks apart at sufficiently high altitudes in the atmosphere. The atoms that make up an O_2 molecule can vibrate about an equilibrium separation, and this motion constitutes a form of internal energy. If the molecule acquires sufficient internal energy, the bond that holds it together can break, resulting in two free oxygen atoms. A specific minimum energy is needed to break the bond, and this amount is much larger than what a molecule might acquire in random collisions with its neighbors. The energy needed to break the molecular bond is called the "binding energy" or the "dissociation energy," and at stratospheric altitudes, a photon of ultraviolet sunlight can provide this. An oxygen molecule absorbs the photon, and the energy of the photon goes into breaking the bond that holds the two atoms together. This photodissociation, introduced previously as R-6, is repeated here:

$$O_2 + h\nu \rightarrow O + O \qquad \text{[R-6]}$$

The frequency ν of the photon in R-6 is related to wavelength λ defined in Chapter 2 by $\nu = c/\lambda$, where c is the speed of light. Recall that the energy of a photon is inversely related related to its wavelength. To dissociate one O_2 molecule, the photon must correspond to a wavelength shorter than 242 nm. This "dissociation wavelength" is far into the ultraviolet portion of the spectrum.

Based on Section 5.1, it is possible to write a quantitative expression for the production rate of oxygen atoms in process R-6. The number of oxygen atoms produced per unit volume of air per unit time at altitude z is denoted by $P(z)$, where typical units are $cm^{-3} s^{-1}$, read as "number of oxygen atoms per cubic centimeter of air per second." Straightforward physical reasoning says that $P(z)$ should increase if solar radiation at the required wavelengths becomes more intense or if the number density of O_2 increases. The production rate is

$$P(z) = 2 J_6(z) [O_2(z)] \qquad \text{[5.4.1]}$$

where $J_6(z)$ is the dissociation rate of O_2 at altitude z, expressed in s^{-1}, and $[O_2(z)]$ is the number density of O_2 in cm^{-3}. The factor of 2 indicates that each time R-6 takes place, two atoms of oxygen are created. With reference to reaction R-2 and Eq. 5.1.1, both products in R-6 are the same atom, and the production rate in Eq. 5.4.1 includes both of these. Based on Eq. 5.2.9 the dissociation rate depends on details of the molecular structure of O_2 and on the intensity of ultraviolet sunlight at wavelengths less than 242 nm. The abundance of O_2 as a function of altitude is well known; at altitudes below 90 km, it makes up 21% of the total atmospheric number density. The fraction of the total O_2 amount that is broken apart in R-6 is extremely small here, although at still greater heights this is no longer true.

At what altitudes in the atmosphere are the free oxygen atoms produced? Are they created at the ground, or at very great heights, or somewhere in between? The answer depends on how both $J_6(z)$ and $[O_2(z)]$ vary with altitude. From hydrostatic balance, the number density of O_2 decreases exponentially as altitude increases. As a reasonable approximation, $[O_2(z)]$ drops by a factor of two for each 5-km increase in altitude from the ground up to around 90 km. The left panel of Fig. 5.3 illustrates this behavior schematically. The dissociation rate has a very different altitude dependence. The quantity $J_6(z)$ varies with altitude because the flux of ultraviolet sunlight changes dramatically depending on location in the vertical. The sun supplies a certain number of photons per unit time at the top of the atmosphere in the wavelength range shorter than 242 nm. As these photons move downward, molecular oxygen absorbs some of them at each altitude along the way. Every time a dissociation occurs, there is one less photon remaining to move to lower altitudes. The result is that $J_6(z)$ shrinks as altitude decreases, as depicted in the middle panel of Fig. 5.3. At high altitudes, above about 80 km, the flux of solar photons at the required ultraviolet wavelengths is large. However, as the beam of sunlight moves downward, it en-

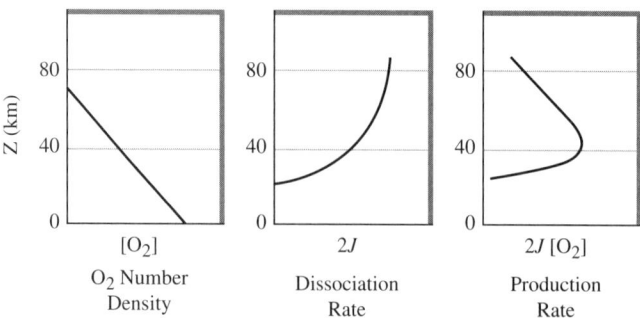

FIGURE 5.3 The altitude dependence of the atomic oxygen production rate in $O_2 + h\nu \rightarrow O + O$ (right side) depends on the vertical profile of O_2 (left side) and the dissociation rate (center). [Adapted from Goody and Walker (1972).]

counters ever increasing amounts of O_2, so the loss of photons per unit change in altitude becomes greater. By an altitude of 20 to 25 km, essentially all of the photons that are energetically able to break an O_2 molecule apart have been absorbed and, consequently, the dissociation effectively shuts off at lower heights.

The production rate of oxygen atoms depends on the product of the two leftmost curves in Fig. 5.3, and this is depicted by the right-hand curve. Note that the dissociation rate is multiplied by a factor of 2 since each dissociation creates two oxygen atoms. The product of $J_6(z)$ and $[O_2(z)]$ leads to a layer with a characteristic shape. At the highest altitudes in Fig. 5.3, where the flux of ultraviolet sunlight has experienced little attenuation, the production rate increases downward, as does $[O_2(z)]$. Eventually, however, absorption leads to a substantial depletion of the radiant flux at wavelengths less than 242 nm. At this point, the production rate undergoes a maximum, bends over, and then shrinks rapidly. The production rate is negligible when very few photons with sufficient energy to break the bond in O_2 survive to reach the altitude in question. The result is a well-defined layer where oxygen atoms are produced in the greatest amounts. The maximum production rate occurs near a height of 40 km, and the production is insignificant at altitudes less than about 20 km. As described in the following, the oxygen atoms released in the photodissociation of O_2 go on to form ozone. The conclusion here as follows: The Earth's ozone layer is located in the stratosphere because this is where most of the oxygen atoms are produced in the photodissociation of molecular oxygen.

The oxygen atoms formed in process R-6 are chemically active. Some of these atoms collide with O_2 molecules, and this is a frequent event since 21% of the atmosphere consists of O_2. In some of these collisions, the O and O_2 create ozone in a three-body reaction:

$$O + O_2 + M \rightarrow O_3 + M \qquad \text{[R-7]}$$

Any given ozone molecule lives only a short time because it can absorb photons across a broad range of the sun's ultraviolet spectrum and undergoes photodissociation. Although ozone absorbs sunlight at ultraviolet wavelengths as long as 340 to 350 nm as well as in the visible, the absorption does not become efficient until wavelength shrinks below 320 nm. The photodissociation is

$$O_3 + h\nu \rightarrow O_2 + O \qquad \text{[R-8]}$$

Process R-8 is responsible for the sharp drop-off in solar energy received at the ground, shown in Fig. 5.2, and this absorption shields the surface of the Earth from exposure to a biologically harmful component of the sun's emission.

Chapman's original theory included one more chemical reaction of relevance here. At this point the atmosphere contains both O and O_3. On occasion these constituents collide with each other and react. This is

$$O + O_3 \rightarrow O_2 + O_2 \qquad \text{[R-9]}$$

Here both of the chemically active trace gases, O and O_3, are converted back to O_2, the long-lived form of atmospheric oxygen. The four reactions, R-6 through R-9, describe the formation of an ozone layer in a pure oxygen atmosphere, where this terminology refers to the fact that the theory includes only oxygen atoms in various combinations. Figure 5.4 depicts the chemistry in the form of a diagram. Each chemical constituent is written inside a circle. Arrows pointing into a circle indicate production of that atom or molecule, and arrows pointing out of a circle indicate loss.

As an example of how to develop a quantitative model, consider reaction R-7. A mixture of O and O_2 leads to formation of O_3. The production rate of ozone is

$$\{d[O_3]/dt\}_7 = k_7[M][O_2][O] \qquad \text{[5.4.2]}$$

where the subscripts identify the reaction number. The production rate of ozone in R-7 depends on the product of all of the reactant abundances, whose number densities are [O], $[O_2]$, and [M], where by convention [M] refers to the total number density of the atmosphere, labeled n in earlier chapters. The same type of reasoning applies to loss rates as well. For example, ozone is destroyed by dissociation in R-8, and the associated loss rate is

$$\{-d[O_3]/dt\}_8 = J_8[O_3] \qquad \text{[5.4.3]}$$

where the minus sign indicates that the process causes a decrease in the number density of ozone over time. Production and loss rates provide the

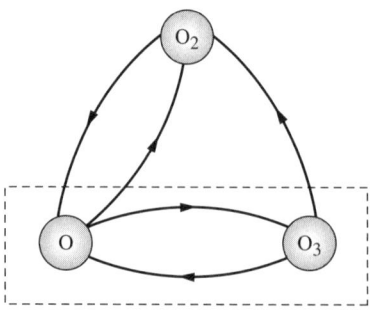

FIGURE 5.4   The chemistry of a pure oxygen atmosphere. Arrows going toward a constituent indicate production. Reactions that convert O into O_3 and back again are very rapid compared to those that link O and O_3 to O_2.

means to convert a list of chemical reactions into a quantitative theory to compute the abundance of the trace gases involved.

The outlined methodology, combined with the chemistry of a pure oxygen atmosphere, allows a calculation of the number density of ozone as a function of altitude. Based on reactions R-6 through R-9, the "continuity equations" for the number densities of atomic oxygen and ozone are

$$d[O]/dt = 2J_6[O_2] + J_8[O_3] - k_7[M][O_2][O] - k_9[O_3][O] \quad [5.4.4]$$

$$d[O_3]/dt = k_7[M][O_2][O] - J_8[O_3] - k_9[O_3][O] \quad [5.4.5]$$

where the right-hand side of each equation is the sum of all individual production and loss rates for each species, with account taken of the sign. In the state of "chemical equilibrium" the number densities of atomic oxygen and ozone are constant in time, that is, $d[O]/dt = d[O_3]/dt = 0$, or the total production rate balances the total loss rate. This is an approximation because the changing location of the sun in the sky over a day and the sequence of daylight and darkness lead to time-dependent changes in the dissociation rates J_6 and J_8. However, for simplicity, this example assumes chemical equilibrium, so that

$$0 = 2J_6[O_2] + J_8[O_3] - k_7[M][O_2][O] - k_9[O_3][O] \quad [5.4.6]$$

$$0 = k_7[M][O_2][O] - J_8[O_3] - k_9[O_3][O] \quad [5.4.7]$$

In principle, one can obtain algebraic solutions to Eqs. 5.4.6 and 5.4.7 for [O] and [O$_3$]. Some simplifications, which are justified by the magnitudes of the various terms, make this straightforward. It is an observed fact that the reactions that convert O into O$_3$ and O$_3$ back into O, being R-7 and R8, are very rapid compared to R-6 and R-9. To a very good approximation, it is true that

$$0 = k_7[M][O_2][O] - J_8[O_3] \quad [5.4.8]$$

Eq. 5.4.8 specifies that ratio of [O] to [O$_3$] in terms of known rate coefficients and the number densities of major atmospheric gases, [M] and [O$_2$]. Two equations are necessary to solve for the two unknowns, [O] and [O$_3$]. To obtain a second equation, addition of Eqs. 5.4.6 and 5.4.7 produces an exact relationship:

$$0 = 2J_6[O_2] - 2k_9[O_3][O] \quad [5.4.9]$$

Eqs. 5.4.8 and 5.4.9 are readily solved to give the number density of ozone at any altitude in the atmosphere. This is

$$[O_3] = \{J_6 k_7 [M][O_2]^2/(J_8 k_9)\}^{1/2} \quad [5.4.10]$$

where all of the quantities on the right-hand side of Eq. 5.4.10 are known as functions of altitude. The dissociation rate of ozone, J_8, involves longer wavelengths and has a much weaker dependence on altitude than does J_6. Hence, the ratio J_6/J_8 in Eq. 5.4.10 decreases as altitude decreases.

The two chemically active trace gases, atomic oxygen and ozone, are referred to collectively as "odd oxygen," indicating that they consist of an odd number of atoms. In this terminology the long-lived molecule, O_2, is "even oxygen." The rapid reactions R-7 and R-8 convert one member of the odd oxygen family into another, meaning that one and only one member of the odd oxygen family appears on each side of the reactions. Reactions R-6 and R-9 are the "source" and "sink" of odd oxygen, respectively, meaning that one even oxygen appears on one side of the reaction and two odd oxygens appear on the other side. It is often convenient to use the term *odd oxygen* without identifying a specific member of the family. This is valid because one member of the odd oxygen lives only a short time before being converted to the other in R-7 or R-8.

This derivation assumes that chemical production and loss are the only mechanisms that determine the abundance of odd oxygen at any point in space. Although this is accurate at altitudes above about 35 km, it is not the case throughout the lower stratosphere. Once odd oxygen is created, it can move about with atmospheric winds, and a complete theory of the ozone layer must account for this transport. At altitudes where the chemical loss of odd oxygen is relatively inefficient, atmospheric transport processes exert a major influence on the distribution in both altitude and latitude. Although the production rate of odd oxygen is a maximum in the upper stratosphere, the maximum number density occurs much lower, near 20 to 25 km high. This is the case because vertical motions move ozone to these lower altitudes, where its chemical destruction is slow. Brasseur and Solomon (1984) give a more detailed discussion of the roles of chemistry and transport in controlling the stratospheric ozone distribution.

Figure 5.5 illustrates a typical ozone profile, expressed as number density versus altitude, for middle latitudes. Through the mesosphere and upper stratosphere, the number density increases rapidly as altitude decreases. This behavior reflects the term $[M][O_2]^2$ in Eq. 5.4.10. The dissociation rate of ozone, J_8, in the denominator of Eq. 5.4.10 involves longer wavelengths than does J_6. Unlike J_6, J_8 has relatively little altitude dependence, so that it has a minor effect on the shape of the ozone profile. When the production rate of oxygen atoms undegoes its maximum and then shrinks with decreasing altitude, the effect of chemistry alone is to make $[O_3]$ decrease as well, unless another process, specifically transport, acts counter to this. At altitudes below 35 km, the action of atmospheric motions invalidates Eq. 5.4.10, and far more ozone resides in the lower stratosphere than predicted by a pure chemical theory.

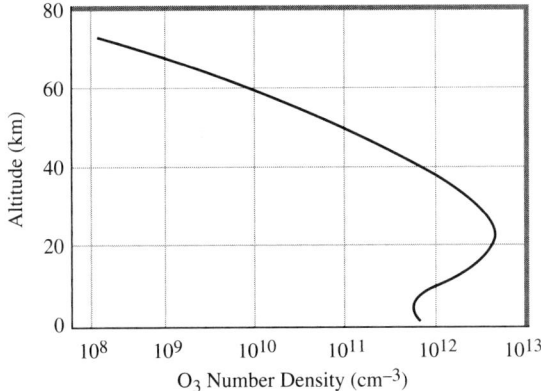

FIGURE 5.5 A typical vertical profile of atmospheric ozone, expressed as ozone number density as a function of altitude *(U.S. Standard Atmosphere, 1976).*

In 1930, when Chapman proposed the first theory of the ozone layer, definitive information was not available to specify all of the reaction rate coefficients or the ultraviolet solar flux required to compute dissociation rates. The theory was successful in that, first, it predicted the existence of an ozone layer and, second, it placed most of the ozone at high altitudes. Given the scarcity of observations, this is where the theory remained for several decades. The next major advances came in the development of rocket-based techniques to measure the vertical distribution of ozone through the stratosphere and the ultraviolet solar flux incident from above. Craig (1965) gives further information on these early investigations. In addition, laboratory studies provided values for the important rate coefficients. This improved knowledge produced an unexpected result: calculations of the ozone abundance in the upper stratosphere based on the updated information predicted larger number densities than were actually observed. From the persistent discrepancy between atmospheric observations and calculations, it became apparent that mechanisms exist for the destruction of odd oxygen in addition to reaction R-9 in Chapman's theory. The solution to this problem led to the realization that human activities could influence the amount of ozone in the stratosphere.

5.5 Chemical Processes Involving Chlorine

Refrigeration became common during the first half of the twentieth century, and a class of chemicals called "chlorofluorocarbons" or "CFCs" were of great value in this important technology. Another widespread use of these chemicals was as propellants in spray cans. The name *CFC* refers to a family of molecules that consist of chlorine, fluorine, and carbon, where the two most abundant members of the group have the chemical symbols CF_2Cl_2 and $CFCl_3$. When locked in a refrigerator or air conditioner, CFCs are liquids, but in practice some of these molecules leak into the atmosphere, where they vaporize to become trace gases. From the 1950s through the early 1990s, the abundance of CFCs in the atmosphere grew at a rate of several percent per year. At their maximum near the year 2000, approximately 8 of every 10 billion air molecules were CFCs, giving a mixing ratio of 8×10^{-10} (World Meteorological Organization 2003). The term "mixing ratio," introduced in Chapter 3 for water vapor, refers to the number density of a trace gas divided by the total number density of the atmosphere at the same location.

The lifetimes of CFCs in the atmosphere are quite long, so once released into the air these molecules remain there for decades. Most CFCs entered the atmosphere in the middle latitudes of the Northern Hemisphere, but winds transported these gases over the entire globe, both horizontally and vertically, so that the mixing ratio became nearly constant with latitude and altitude up to approximately 30 km above the ground. Once in the upper stratosphere, the CFCs are exposed to portions of the sun's ultraviolet radiation, and photons that correspond to wavelengths less than about 220 nm can break chlorine atoms from the molecules:

$$CFCl_3 + hv \rightarrow CFCl_2 + Cl \qquad \text{[R-10]}$$

$$CF_2Cl_2 + hv \rightarrow CF_2Cl + Cl \qquad \text{[R-11]}$$

The CFCs must drift to altitudes above 30 km before encountering sufficiently short solar wavelengths because of the strong attenuation provided by O_2 in this spectral region. As a consequence, processes R-10 and R-11 cannot operate in the lower stratosphere because of the absence of photons with the required energies. The important outcome of reactions R-10 and R-11 is the presence of free chlorine atoms in the upper stratosphere.

Chlorine atoms destroy odd oxygen through an efficient sequence of reactions. The first step is

$$Cl + O_3 \rightarrow ClO + O_2 \qquad \text{[R-12]}$$

The chlorine monoxide, ClO, then reacts with an oxygen atom:

$$ClO + O \rightarrow Cl + O_2 \qquad \text{[R-13]}$$

Reactions R-12 and R-13 each destroy one member of the odd oxygen family and produce one even oxygen. It is possible to add chemical reactions together just as one adds equations. The sum of R-12 and R-13 is

$$Cl + O_3 + ClO + O \rightarrow ClO + O_2 + Cl + O_2 \qquad \text{[R-14]}$$

As with equations, one can cancel a term when it appears on both sides of the expression. In R-14 both Cl and ClO cancel to give the net reaction:

$$O_3 + O \rightarrow O_2 + O_2 \qquad \text{[R-15]}$$

This is identical to R-9 in the chemistry of a pure oxygen atmophere, so the net effect of R-12 and R-13 operating together is the same as if reaction R-9 proceeds at a faster rate. Molina and Rowland (1974) first proposed that atmospheric chlorine can lead to the destruction of ozone and thereby launched an era of intense research into the chemistry of the ozone layer.

Reactions R-12 and R-13 make up a "catalytic cycle." A catalytic cycle is a series of reactions in which the atom or molecule that initiates the sequence is recreated at the end. In this case, the chlorine atom that initiates R-12 is recreated in R-13, so it is available to go through the cycle numerous times until some other process removes the atom. Atomic chlorine is the "catalyst" for the cycle. A catalytic cycle allows a very small amount of chlorine to destroy a large amount of ozone, and this is why the growing atmospheric abundance of the CFCs caused so much concern. In the absence of leakage of CFCs from cooling equipment, natural sources would provide very little chlorine to the stratosphere. Throughout the second half of the twentieth century, the CFC abundance in the atmosphere rose rapidly as refrigeration and air conditioning were in high demand. Consequently, there was more Cl in the stratosphere each year than the year before, and the ozone abundance began to decline in response to the catalytic cycle.

Chlorine atoms are removed from the atmosphere by a combination of chemical processes, transport, and, ultimately, the formation of precipitation in the troposphere. One of the trace gases involved here is methane, CH_4. Various biological processes create methane at the Earth's surface, and its lifetime is sufficiently long that motions carry some of it into the stratosphere. Chlorine atoms react with methane to produce hydrochloric acid, HCl, as follows:

$$Cl + CH_4 \rightarrow CH_3 + HCl \qquad \text{[R-16]}$$

The formation of HCl is important for two reasons. First, reaction R-16 takes chlorine atoms out of the ozone destruction cycle; without the formation of HCl, the loss of ozone would have been much larger than actually observed. Second, HCl lives sufficiently long that atmospheric motions can transport it long distances, including downward into the troposphere. Once in the troposphere, the availability of liquid water in clouds is critical. The hydrogen atom makes the HCl molecule soluble in water. Consequently, hydrochloric acid becomes incorporated into cloud droplets and eventually falls out in precipitation. This "rainout" is an efficient way to remove pollutants from the atmosphere.

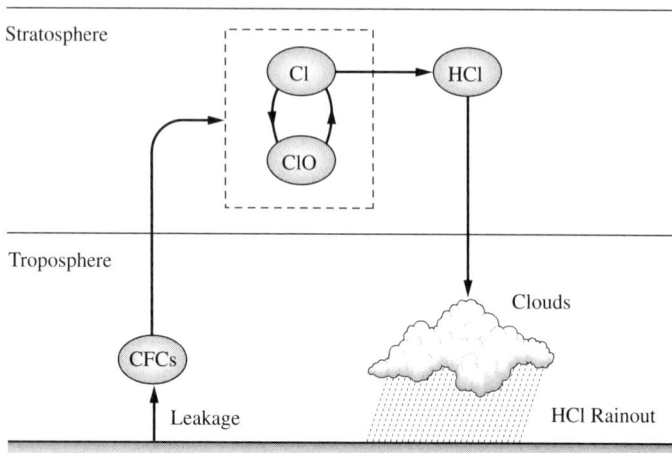

FIGURE 5.6   Cycling of chlorine through the atmosphere. During the twentieth century, chlorine bound in CFCs leaked into the atmosphere. Ultraviolet sunlight liberated chlorine atoms in the upper stratosphere. Eventually the chlorine forms hydrochloric acid, which falls out in precipitation.

Figure 5.6 illustrates the cycling of chlorine through the atmosphere. Chlorine enters the atmosphere at the ground bound in chlorofluorocarbons. Vertical motions carry some of the CFCs into the upper stratosphere, where ultraviolet sunlight releases free chlorine atoms, followed by a catalytic cycle that destroys odd oxygen. Eventually, hydrochloric acid forms, followed by downward transport and rainout. The time required for a chlorine atom to complete the entire cycle in Fig. 5.6 is measured in decades to a century. Concern about the chemical effects of CFCs in the stratosphere prompted development of alternative chemicals for use in refrigeration, and today the manufacture of CFCs has ceased. In the coming decades the chlorine content of the stratosphere should decline, and the stratospheric ozone abundance should gradually return to its unperturbed value.

5.6   The Excited Oxygen Atom, $O(^1D)$

The electrons that surround an atomic nucleus can reside in several different orbits, where each arrangement corresponds to a different energy state of the atom. When ozone photodissociates in reaction R-8, the resulting oxygen atom might be produced in two different states. The lowest energy state, called the ground state, is labeled $O(^3P)$, pronounced "O triplet P." The next excited state is $O(^1D)$, pronounced "O singlet D." Laboratory measurements show that when ozone absorbs light at wavelengths shorter than 313 nm, the oxygen atom produced is most likely to be in the ^1D excited state, whereas wavelengths longer than 313 nm yield the ^3P state. The processes are

$$O_3 + hv \rightarrow O_2 + O(^1D) \text{ for } \lambda < 313 \text{ nm} \qquad \text{[R-17]}$$

$$O_3 + hv \rightarrow O_2 + O(^3P) \text{ for } \lambda > 313 \text{ nm} \qquad \text{[R-18]}$$

For ease in notation, the symbol O will refer to the ground state $O(^3P)$. The important point for subsequent chemistry is that $O(^1D)$ can participate in chemical reactions that are not energetically possible for $O(^3P)$.

5.7   Chemical Processes Involving Nitrogen

A catalytic cycle similar to that described in Section 5.5 takes place with trace gases that contain nitrogen. Unlike the case with chlorine, the catalytic cycle associated with nitrogen would be an important influence on stratospheric ozone even in the absence of human activity. Organisms in the soil and ocean produce the trace gas nitrous oxide, N_2O, and because of a long lifetime in the atmosphere, motions mix it horizontally and vertically over the globe. The mixing ratio of N_2O in the troposphere and lower stratosphere is approximately 3×10^{-7}, meaning that there are 3 molecules of N_2O for every 10^7 molecules of air. A reaction with excited oxygen atoms breaks the bond in N_2O and produces two molecules of nitric oxide, NO:

$$O(^1D) + N_2O \rightarrow NO + NO \qquad \text{[R-19]}$$

Nitric oxide then destroys ozone and atomic oxygen in a catalytic cycle that is analogous to that involving chlorine:

$$NO + O_3 \rightarrow NO_2 + O_2 \qquad \text{[R-20]}$$

$$NO_2 + O \rightarrow NO + O_2 \qquad \text{[R-21]}$$

where the molecule NO_2 is "nitrogen dioxide" (Crutzen 1970, 1971). The net reaction is $O + O_3 \rightarrow O_2 + O_2$, the same as in the chlorine catalytic cycle, and the effect is as if the sink of odd oxygen in the original theory of the ozone layer, R-9, operates at a greatly increased rate. One molecule of nitric oxide can go through the catalytic cycle numereous times, thereby destroying a substantial amount of ozone.

The nitrogen catalytic cycle provides a means for high-flying aircraft to influence the atmospheric ozone abundance, since nitric oxide is present in the exhaust. If this NO is released into the troposphere, where most aircraft operate, it is chemically converted into water-soluble molecules, which fall out in precipitation without reaching stratospheric altitudes. However, NO introduced directly into the stratosphere can begin destroying ozone via R-20 and R-21. This issue arose in the early 1970s, when plans were underway to build a fleet of supersonic passenger aircraft designed to fly in the lower stratosphere. If technological improvements in aircraft engines can limit the NO released to small quantities, this will not be a major issue in the future, when such aircraft may operate routinely.

The NO and NO_2 must eventually exit the atmosphere, and the water molecule is critical here. The discussion of saturation in Sections 3.3 and 3.5 implies that very little water vapor can survive to reach the stratosphere. Water vapor moves from troposphere to stratosphere through the upward branch of the Hadley circulation in the tropics. As a consequence of this trajectory, the water vapor abundance in the stratosphere is controlled by the saturation vapor pressure at the coldest point along the path, the tropical tropopause. This leads to a mixing ratio of 4 to 5×10^{-6}, so that of every 1 million air molecules in the stratosphere, only 4 to 5 are H_2O.

The excited oxygen atom breaks the bond in H_2O to yield two hydroxyl radicals, OH:

$$O(^1D) + H_2O \rightarrow OH + OH \qquad \text{[R-22]}$$

The hydroxyl radical then reacts with NO_2 in a three-body process to create nitric acid, HNO_3:

$$OH + NO_2 + M \rightarrow HNO_3 + M \qquad \text{[R-23]}$$

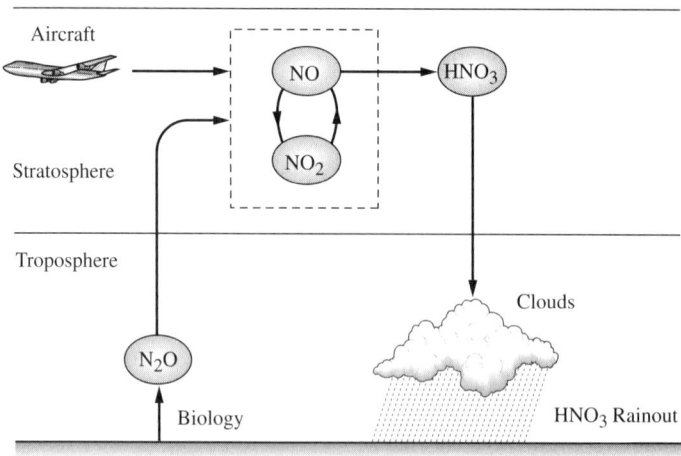

FIGURE 5.7   Cycling of chemically active nitrogen through the atmosphere. Nitrous oxide drifts to stratospheric altitudes, where reaction with $O(^1D)$ releases nitric oxide. High-flying aircraft can inject nitric oxide directly into the stratosphere. Eventually nitric acid forms and falls out in precipitation.

Nitric acid behaves similarly to hydrochloric acid in Section 5.5. Nitric acid has a sufficiently long atmospheric lifetime that motions transport it to the troposphere, where the molecule becomes incorporated into cloud droplets and rains out. Figure 5.7 summarizes the cycling of chemically active atmospheric nitrogen. Note the similarity to the cycling of chlorine.

5.8 The Antarctic Ozone Hole

The final topic related to stratospheric ozone concerns the large depletion that occurs each spring at high latitudes in the Southern Hemisphere, called the Antarctic ozone hole. To place the discovery of the ozone hole in perspective, it is useful to review the theoretical understanding of stratospheric ozone depletion as of 1985. At that time it was widely accepted that the increasing abundance of CFCs would be accompanied by a decline in the concentration of ozone in the stratosphere. Chemical models at the time predicted that the largest percentage change in ozone number density would occur in the upper stratosphere, near an altitude of 40 km. Fortunately, a change in the local ozone amounts at these altitudes would have only a small impact on the total number of ozone molecules in a column of the atmosphere, since most of the ozone resides lower down. Inspection of Fig. 5.5 shows that roughly 80% of the ozone molecules in a vertical atmospheric column lie at altitudes below 30 km. This is the result of downward transport of odd oxygen from the upper stratosphere, where it is produced most efficiently, into the lower stratosphere, where loss is slow. When the concern is with effects of increased ultraviolet sunlight at the ground, the relevant quantity is the total column ozone amount. Although the predicted percentage declines in local ozone amounts near 40 km were substantial, the resulting increases in ultraviolet exposure at the ground would have been small.

Why did the predicted change in local ozone amounts maximize near 40 km instead of at a lower altitude? A major part of the answer appeared in Section 5.5. Chlorine atoms are responsible for the catalytic cyle that destroys odd oxygen. However, in the lower stratosphere most of the Cl is converted to hydrochloric acid, and when this occurs, the catalytic cycle is no longer effective. In addition to HCl, the molecule chlorine nitrate, $ClONO_2$, forms in a three-body reaction between ClO and NO_2. Thanks to the formation of HCl and $ClONO_2$, the theory said that ozone at altitudes below about 30 km was less strongly influenced by the release of CFCs than was the case higher up. The ozone destruction became very inefficient at the altitudes where ozone was most abundant. By early 1985, there was wide recognition that some degree of ozone loss was occurring, but the prevailing chemical theory indicated that this would be small and spread over decades.

Observations made in Antactica, reported by the British Antarctic Survey in 1985 (Farman et al. 1985), demonstrated that this theory was not complete. The research group had performed measurements of total column ozone from a scientific base at Halley Bay, latitude 76° south, since 1957. The observations utilize ultraviolet sunlight reaching the Antarctic surface at various wavelengths, so no data are available during the period of winter darkness. Every September, when sunlight returned with the approach of spring, the column ozone measurements began. This is when a pattern of unusual behavior appeared and evolved over the years of observation. From 1957 to about 1970 the column ozone amount displayed essentially normal behavior from September to December, but in the early 1970s a dip became noticeable in late September. Minimum ozone amounts existed in October, followed by a recovery to normal levels in November or, at the latest, early December. This behavior was unexpected, but the most unusual aspect was that the minimum ozone amount tended to become smaller from one year to the next.

During the early years, the minimum in October was not very obvious and did not attract much attention. There is always a degree of random year-to-year variability in column ozone over Antarctica, and a few years of abnormally low amounts did not raise great concern. However, as the years progressed, it became apparent that springtime ozone amounts were experiencing a major decline. Between 1975 and 1985 the downward trend became more pronounced, with column ozone amounts in the Octobers of the mid 1980s falling to around 60% of the unperturbed values of the 1960s. This springtime decrease in column ozone became known as the Antarctic "ozone hole."

The large depletion in high-latitude ozone appears only in the spring, and ozone amounts return to normal by December or earlier. Furthermore, the geographic extent of the ozone depletion is limited. Its boundaries are usually confined to austral latitudes south of 60°, although on occasion areas of depleted ozone can extend over the southern tip of South America. During the 1980s and 1990s, the smallest recorded ozone amounts tended to shrink, as depicted in Fig. 5.8, which presents the minimum column ozone abundance observed anywhere over Antarctica in each year based on satellite-borne sensors operated by the National Aeronautucs and Space Administration. Although these are the extremes, reductions in column ozone of 50% are common over large areas of the Antarctic continent.

The ozone hole is an unexpected consequence of introducing chlorine into the stratosphere, where some chemistry peculiar to the polar regions occurs. Solomon (1999) has presented a thorough discussion of the processes involved, and the World Meterological Organization (2003) assessment contains an account of the available observations. As noted, most of the ozone molecules in a vertical column of the atmosphere are located at altitudes below 30 km, and at these altitudes the chlorine atoms become bound into the molecules $ClONO_2$ and HCl, at least according to the theory that existed in 1985. When this is the case, there is little destruction of ozone. Over Antarctica, however, a different situation prevails during spingtime. When sunlight

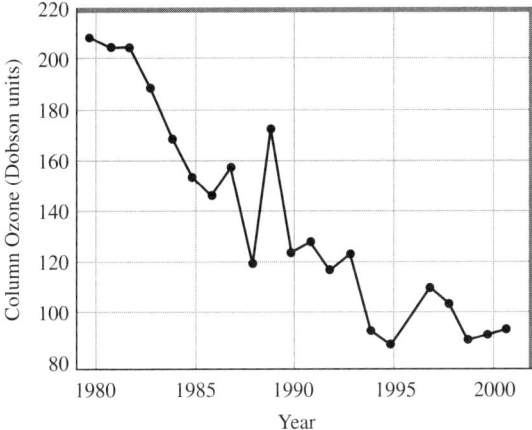

FIGURE 5.8   Minimum column ozone amount observed over Antarctica each spring for the years 1980–2001. Values are in "Dobson units," where 1 Dobson unit equals 2.69 $\times$ 10^{20} ozone molecules in a vertical column with a 1 m^2 horizontal area. (Source: NASA Goddard Space Flight Center)

returns to Antarctica in September, some unique chemical processes, which operate only in very low temperatures, liberate free chlorine atoms in the lower stratosphere, the altitude region where most of the ozone resides. The result is a substantial catalytic destruction of column ozone during the spring season.

There are several important questions. Why does this peculiar ozone depletion exist only over Antarctica? Why does it appear in September and vanish by November or December? Finally, why did the ozone loss become increasingly pronounced in the years after the mid 1970s? The origin of the ozone hole requires assembling several pieces of information involving motions in the high-latitude stratosphere, solar radiation, the properties of water, and atmospheric chemistry. The first part of the explanation involves wind systems. When winter develops over Antarctica, strong stratospheric winds circulate above the continent. This swirling of air is the "polar vortex" in which wind blows around the continent, but not southward over it. The result is that essentially the same mass of air remains over Antarctica for the winter season, during which time the transport of warmer air from lower latitudes does not occur. Heating from sunlight is absent during the polar winter, while the air continues to emit longwave radiation to become ever colder. The stratosphere over Antarctica in winter is colder than that in the Arctic because poleward heat transport is less efficient in the Southern Hemisphere. During winter, Antarctica tends to be thermally isolated from middle latitudes, and the result is an extremely low stratospheric temperature. This very cold air is an essential ingredient in the formation of the ozone hole.

Chapter 3 describes changes in phase of water. Very little water vapor exists at stratospheric altitudes because most of it has condensed to form clouds in the troposphere. As noted in Section 5.7, the stratospheric mixing ratio of water vapor is only 4 to 5×10^{-6}. However, in the very cold Antarctic lower stratosphere, even this small quantity of water can exceed the saturation vapor pressure, which from Eq. 3.5.1 has a very strong dependence on temperature. The result is that during winter, water vapor freezes to form tiny ice crystals in the lower stratosphere over Antarctica. Collections of these ice crystals constitute "polar stratospheric clouds" (PSCs). In addition, nitric acid, HNO_3, becomes incorporated into the ice, and HCl molecules stick to the surfaces of the crystals.

The next step in the sequence is critical. During the winter darkness, when PSCs are present, chemical reactions take place on the surfaces of the ice crystals. These reactions would not occur if two gas-phase molecules collided; the presence of a solid surface is necessary for the processes to proceed. The result is that hydrochloric acid and chlorine nitrate are destroyed, and new chlorine-containing molecules are created. The two most important reactions are:

$$ClONO_2 + H_2O(\text{solid}) \rightarrow HOCl + HNO_3 \qquad [\text{R-24}]$$

$$HCl + HCl(\text{solid}) \rightarrow Cl_2 + H_2 \qquad [\text{R-25}]$$

where the addition of the term "solid" on the reactants indicates that the molecule resides on the surface of an ice crystal. During the polar winter in the presence of PSCs, the abundances of $ClONO_2$ and HCl decline, to be replaced by Cl_2 and HOCl. These molecules are weakly bound; little energy is required to liberate free chlorine atoms. When sunlight returns to the polar regions in September, photons in the visible portion of the spectrum lead to dissociation:

$$HOCl + h\nu \rightarrow OH + Cl \qquad [\text{R-26}]$$

$$Cl_2 + h\nu \rightarrow Cl + Cl \qquad [\text{R-27}]$$

The result is a release of free chlorine atoms into the polar lower stratosphere, the altitude region where most of the ozone in a vertical column resides. At this stage a catalytic cycle that destroys ozone can begin. The catalytic cycle over Antarctica is different from the one that operates at higher altitudes, R-12 and R-13, because the atomic oxygen abundance in the lower stratosphere is extremely small, a prediction of Eq. 5.4.8. However, the result is the same, namely, a greatly accelerated chemical removal of ozone. The unique aspect of the polar ozone depletion is that it occurs in the altitude range where ozone is most abundant, and it leads to a large change in the total number of ozone molecules in a column of the atmosphere. These mechanisms explain the development of the ozone hole in September and October.

As time progresses through November, the Antarctic stratosphere warms, and the PSCs evaporate as the saturation vapor pressure increases. Without ice crystals present, reactions R-24 and R-25 no longer operate, chlorine atoms again become incorporated into $ClONO_2$ and HCl, and the ozone destruction ceases. Another important effect involves the change in high-latitude wind systems. As spring proceeds, the polar vortex breaks down, allowing warmer air from lower latitudes to flow over Antarctica. This air both warms the lower stratosphere and brings unperturbed ozone amounts with it. The result is a filling in of the ozone hole. The disintegration of the polar vortex each spring marks the end of that season's ozone depletion. Depending on the exact timing of the vortex's breakup, the ozone hole may dissipate in November or perhaps in early December.

Over the years from 1970 to 1990, the chlorine content of the atmosphere rose at a rate of 5 to 6% annually, and the result was a growing ozone loss over time. Interannual variability in the degree of ozone destruction arises from the meteorology of the lower stratosphere. An unusually warm year will have relatively few PSCs and less ozone loss. The same chemical processes that operate over Antarctica occur in the Northern Hemisphere's polar regions as well. However, the Arctic regions experience an earlier spring warming than does Antarctica, and this leads to much less dramatic ozone depletions in the North.

5.9   Chemical Processes in the Troposphere

Earlier sections of this chapter addressed some important chemical processes that occur in the mesosphere and stratosphere. A key step here was the photodissociation of molecular oxygen, R-6, a reaction that occurs only at middle stratospheric altitudes and above. Indeed, in the absence of R-6 there would be no ozone layer, no $O(^1D)$, and consequently no chemical breakup of N_2O in R-19 and H_2O in R-22. The most important chemistry of the stratosphere would essentially shut down. Given that ultraviolet radiation of the wavelengths required to dissociate O_2 does not reach the troposphere, one might expect little chemical activity in the lower atmosphere. In fact, quite the opposite is true. Myriad biological and anthropogenic processes introduce an array of molecules into the lower atmosphere. Relatively large amounts of water vapor promote production of OH, which, when mixed with gases created at the Earth's surface, initiate a range of chemical activity. The presence of liquid water in clouds ultimately provides a sink via rainout for soluble species created by chemistry. Finally, although solar radiation is limited to wavelengths longer than about 300 nm, selected photodissociations still play a major role in controlling the trace gas composition of the troposphere.

Tropospheric chemistry and air pollution are closely related subjects, although degraded air quality was a concern long before there was an understanding of the molecular-scale mechanisms at work. The earliest air pollution likely consisted of smoke from small fires used for heating and cooking. When human settlements consisted of small villages, this was not a major concern, but as larger cities developed, the smoky air became a significant problem. The first well-documented urban air pollution existed in Rome during the first century CE, where a dark cloud hung over the city and was visible for miles around (Landsberg 1981). The burning of coal for heating homes and businesses caused significant air pollution during the Middle Ages, and this persisted into the twentieth century, until natural gas replaced coal as a primary means of heating relatively small spaces. Today the major use of coal is in electrical power generation, thereby restricting the release of smoke to limited geographic areas.

Smoke contains small particles created in the combustion process. When these are released into an atmosphere where the relative humidity is high, liquid water can condense on the particles to create a fog. Appropriately, the word "smog" merges "smoke" and "fog." Today this relatively simple mixture might be referred to as "London-type smog" to distinguish it from the more recent problem of "photochemical smog." Photochemical smog is a mixture of gaseous and particulate products of combustion plus additional trace gases that are produced by chemistry in the atmosphere. These chemical processes involve sunlight and major gases such as molecular oxygen and water vapor that are present in the natural background troposphere. The most undesirable elements of photochemical smog are the chemically produced trace gases, collectively called "secondary pollutants" to distinguish them from "primary pollutants," which are released directly into the atmosphere. Among the secondary pollutants, ozone that forms at the ground in urban areas has been the subject of major scientific study.

5.10 Primary Pollutants in the Troposphere

Historically, concerns over air quality, especially in cities, have been linked to energy use and the accompanying release of primary pollutants. Raw materials such as metals and stone must be extracted from the Earth and processed to create all of the products on which modern society is based, and these activities require the expenditure of energy. Additional energy is consumed to heat and cool living spaces, to operate electrical appliances, and to move objects and people from one place to another. Since the industrial revolution and continuing into the early twenty first century, the burning of fossil fuels has provided the major energy source used by society. Three types of primary pollutants—hydrocarbons, carbon monoxide, and nitrogen oxides—trace their origins to the widespread use of fossil fuels.

Fossil fuels consist of hydrocarbon molecules in solid (coal), liquid (oil), and gaseous (natural gas) forms. Natural gas is the simplest of the hydrocarbons, consisting primarily of methane (CH_4). The next three larger hydrocarbons are ethane (C_2H_6), propane (C_3H_8), and butane (C_4H_{10}). Note that all of these molecules have the general form C_nH_{2n+2}, where $n = 1, 2, \ldots$ is an integer. Still larger hydrocarbons, such as heptane (C_7H_{16}) and octane (C_8H_{18}), exist as liquids and are major components of gasoline. All of these molecules, referred to as "alkanes," consist of a linear chain of carbon atoms, where each interior carbon is bound to two hydrogen atoms and the carbon atom on each end is bound to three hydrogens. The fundamental unit of coal is a much larger hydrocarbon consisting of more than 100 carbon atoms, 90 to 100 hydrogen atoms, and small amounts of oxygen, nitrogen, and sulfur.

Combustion of hydrocarbons consists of a series of high-temperature chemical reactions that release light and heat. To initiate combustion, it is necessary to heat the fuel to a critical "ignition temperature," which typically is in the vicinity of 900 to 1000 K. At this point reactions with oxygen begin, one effect of which is to release additional heat, which both sustains further reactions and provides a source of energy. In a totally efficient combustion, an alkane is broken down to create only carbon dioxide and water. For example, the combustion of octane can be expressed by the net reaction

$$2\ C_8H_{18} + 25\ O_2 \rightarrow 16\ CO_2 + 18\ H_2O + \text{Energy} \qquad \text{[R-28]}$$

Note the resemblance between R-28 and the net reaction for respiration presented in Section 1.2. The production of energy in mammals after consumption of organic molecules has the same character as the burning of fossil fuels. The net reaction R-28 obscures the fact that combustion consists of a large number of individual steps, and given this complexity, it is not necessarily true that the only products will be carbon dioxide and water. A deviation from R-28, or "inefficient" combustion, is the starting point for air pollution.

As a simple example, consider the combustion of methane, as might occur in a furnace that burns natural gas. The net reaction for an efficient process is

$$2\ CH_4 + 4\ O_2 \rightarrow 2\ CO_2 + 4\ H_2O + \text{Energy} \qquad \text{[R-29]}$$

where the reactants and products are multiplied by 2 to facilitate later comparison with R-30. When the combustion is not totally efficient, some carbon monoxide can be produced via

$$2\ CH_4 + 3\ O_2 \rightarrow 2\ CO + 4\ H_2O + \text{Energy} \qquad \text{[R-30]}$$

The result of the inefficiency is release of CO into the atmosphere or in some cases into indoor air, where it can pose a health hazard when present in sufficient amounts. The inefficient combustion of larger hydrocarbons also produces CO, which is why this trace gas is one of those measured in automobile emissions tests. Breathing excessive CO in confined spaces can cause respiratory distress and, in extreme cases, death. While these hazards of poor indoor air quality are relevant to human health, carbon monoxide is important to tropospheric chemistry because it participates in a series of reactions that produce ozone at the ground.

Any of the hydrocarbons identified here can enter the atmosphere as a result of leakage, spillage, or "incomplete combustion." Incomplete combustion is characterized by the presence of hydrocarbons in the exhaust of an internal combustion engine. These hydrocarbons might be molecules of the original fuel, such as octane, or they can be molecular fragments of the original fuel. For example, instead of proceeding as in R-28, the combustion of octane may produce a fragment of the original molecule that escapes into the atmosphere, as in

$$8 \, C_8H_{18} + 87 \, O_2 \rightarrow 4 \, C_2H_5 + 56 \, CO_2 + 62 \, H_2O + \text{Energy} \quad \text{[R-31]}$$

The C_2H_5 is itself a hydrocarbon. The incomplete combustion of liquid and solid fossil fuels can introduce a wide variety of these "cracked hydrocarbons" into the atmosphere. As such, atmospheric hydrocarbons include numerous possible combinations of carbon and hydrogen, ranging from molecules containing only one carbon atom to the larger molecules that made up the original fuel.

The final category of primary pollutants consists of chemically active nitrogen oxides. The oxides of nitrogen, NO and NO_2, are significant in the stratosphere, where they destroy odd oxygen in a catalytic cycle. It may be surprising that these same gases lead to a net production of ozone at the ground. Air that is introduced into the combustion chamber of an engine is subjected to very high temperatures. Under these extreme conditions, molecules of N_2 and O_2 can decompose, and some of the resulting N and O atoms combine to form NO. Geographic regions with a large amount of automobile and truck traffic are prone to elevated levels of NO. In addition, large localized amounts of NO can form in rural areas through the burning of trees and vegetation. Chemical reactions described in the following section convert much of the NO into NO_2. When a large amount of NO_2 exists over a city, it is visible as a brown cloud because nitrogen dioxide absorbs the violet and blue portions of the spectrum, so the light available to the eye takes on a brownish cast.

5.11 The Formation of Ozone in Urban Areas

As noted in Section 5.9, the mechanisms that lead to ozone formation in the stratosphere cannot operate at the ground because of the absence of photons capable of dissociating molecular oxygen. Yet, measurements typically show ozone mixing ratios in rural air in the vicinity of 20 to 30×10^{-9}

(Logan 1985; Warneck 1988). The classical view was that this amount of ozone flowed downward from the stratosphere to the lower troposphere, where it was ultimately consumed at the ground. Although this is correct in that the downflow does indeed occur, it fails to account for the elevated ozone amounts found in many urban centers. Clearly, there must be a chemical mechanism for creating ozone near the Earth's surface, and the primary pollutants identified in Section 5.10 are central to this process.

If oxygen atoms are present, they form ozone in the three-body reaction R-7. The key issue therefore centers on the creation of atomic oxygen near the ground, and this is where oxides of nitrogen enter the chemistry. As stated previously, nitrogen dioxide photodissociates in the presence of visible sunlight and the longest wavelengths of ultraviolet sunlight:

$$NO_2 + h\nu \rightarrow NO + O \qquad \text{[R-32]}$$

The oxygen atom released in R-32 goes on to form ozone in R-7. This is straightforward, but the difficulty is that the vast majority of chemically active nitrogen released in combustion is in the form of NO, not NO_2. Furthermore, the chemical path for converting NO to NO_2 that operates in the stratosphere, R-20, destroys ozone. Near the ground, there must be a mechanism for converting NO to NO_2 without destroying odd oxygen, and this is where carbon monoxide and the hydrocarbons appear.

The simplest mechanism begins with CO. In daylight, R-22 operates at the ground, producing the hydroxyl radical OH. Given both CO and OH, the following sequence of reactions takes place:

$$CO + OH \rightarrow CO_2 + H \qquad \text{[R-33]}$$

$$H + O_2 + M \rightarrow HO_2 + M \qquad \text{[R-34]}$$

$$NO + HO_2 \rightarrow NO_2 + OH \qquad \text{[R-35]}$$

Reaction R-35 provides the NO_2 that photodissociates in R-32 and via R-7 leads to formation of ozone at the ground.

Seinfeld and Pandis (1998) provide a detailed account of the chemistry of atmospheric hydrocarbons. Although the hydrocarbon family consists of a large number of different molecules, their chemical behavior is similar. A common reaction is one in which the hydrocarbon gives up a hydrogen atom. It is convenient to adopt the notation "RH" to refer to all members of the hydrocarbon family taken collectively. If, for example, the hydrocarbon is octane, then the symbol R is equivalent to the molecule C_8H_{17}. To start the chemistry, a hydrocarbon transfers a hydrogen atom to OH to form water vapor:

$$RH + OH \rightarrow R + H_2O \qquad \text{[R-36]}$$

This is followed by the sequence

$$R + O_2 + M \rightarrow RO_2 + M \qquad [R-37]$$

$$RO_2 + NO \rightarrow RO + NO_2 \qquad [R-38]$$

Given the NO_2 from R-38, the creation of ozone is straightforward via R-32 and R-7. Although one ozone molecule is produced via the preceding chemistry, the hydrocarbon RO created in R-38 forms still more ozone. The fragment R of the original hydrocarbon still contains several hydrogen atoms. If this is denoted by R = R*H, then the molecule RO can be written as RO = R*HO. The chemistry continues as follows:

$$R*HO + O_2 \rightarrow R*O + HO_2 \qquad [R-39]$$

The HO_2 molecule produced here creates another NO_2 in R-35 and reactions R-32 and R-7 lead to formation of ozone.

Both reactions R-33 and R-36, which start the sequence leading to ozone production at the ground, involve the hydroxyl radical. The ultimate source of OH is reaction R-22 involving $O(^1D)$, which itself traces back to the photodissociation of ozone at wavelengths less than about 313 nm, R-17. This demonstrates that a small amount of ozone at the ground is necessary to produce additional ozone. As noted previously, this initial ozone can flow downward from the stratosphere. Once some ozone exists near the ground, the sequence that creates still more ozone can proceed.

Winds can transport ozone away from the urban area where it was produced and disperse it over a much larger area. Eventually the ozone is removed from the atmosphere by a combination of chemistry and uptake at the Earth's surface. Chemical removal of ozone by the nitrogen catalytic cycle is efficient in the stratosphere, but a numerical evaluation of Eq. 5.4.8 using observed ozone amounts produces extremely small values of [O] in the lowest few kilometers of the atmosphere. Hence, the loss of odd oxygen near the ground must occur by means other than the combined action of R-20 and R-21. The production of $O(^1D)$ in R-17 followed by R-22 with water vapor, which is abundant in the troposphere, constitutes a substantive chemical removal of ozone. Finally, an important nonchemical loss mechanism for ozone over land involves uptake by plants (Seinfeld and Pandis 1998).

Improvements in engineering have led to greatly decreased NO, CO, and RH emissions from internal combustion engines from those of several decades ago. In response, peak daily ozone mixing ratios in excess of 1.2×10^{-7} are much less frequent today than in the last half of the twentieth century. Because of technological advances, air quality in major American cities has improved considerably, although this is not necessarily the case in a number of developing countries.

5.12 Exercises

1. Assume that the average scale height of the atmosphere at altitudes between 10 and 70 km is $H = 6.9$ km. Use the theory of ozone in a pure oxygen atmosphere to show that the ozone number density can be described by a scale height of approximately 4.6 km.

2. Assume that the following two reactions (R-20 and R-21) alone determine the ratio of the nitric oxide number density to the nitrogen dioxide number density:

$$NO + O_3 \rightarrow NO_2 + O_2 \qquad\qquad [\text{R-20}]$$

$$NO_2 + O \rightarrow NO + O_2 \qquad\qquad [\text{R-21}]$$

where the rate coefficients are k_{20} and k_{21}, respectively. What is the ratio $[NO]/[NO_2]$ expressed in terms of $[O]$ and $[O_3]$ and the rate coefficients?

3. Consider the formation of an ozone layer in a pure-oxygen atmosphere, reactions R-6 through R-9 in the text, but modified by the nitrogen catalytic cycle in Exercise 2 above, R-20 and R-21. Assume chemical equilibrium, assume that Eq. 5.4.8 is valid, and adopt the result of Exercise 2 for the ratio $[NO]/[NO_2]$. Derive an algebraic expression for the number density of ozone when all of these processes operate. (*Note:* A solution to the quadratic equation $ax^2 + bx + c = 0$ is $x = \{-b + [b^2 - 4ac]^{1/2}\}/(2a)$.)

4. A second solution to the quadratic equation in Exercise 3 exists, with a minus sign in front of the square root term. Explain on physical grounds why this solution is not acceptable in Exercise 3.

5. Chemically active forms of nitrogen influence the odd oxygen abundance via reactions R-20 and R-21 (repeated in Exercise 2) and the dissociation of NO_2 in R-32, $NO_2 + h\nu \rightarrow NO + O$ with dissociation rate J_{32}. To a good approximation the production rate of NO_2 in R-20 is balanced by the loss rates of NO_2 in R-21 and R-32 combined. Given this, show that the net loss rate of odd oxygen caused by the action of NO and NO_2 is equal to $2k_{21}[NO_2][O]$.

5.13 References

Brasseur, G., and S. Solomon. 1984. *Aeronomy of the Middle Atmosphere.* Dordrecht, Holland: D. Reidel Publishing Co.

Chapman, S. 1930. A theory of upper atmospheric ozone. *Mem. Royal Meteorol. Soc.* 3:103–25.

Craig, R. A. 1965. *The Upper Atmosphere: Meteorology and Physics.* New York: Academic Press.

Crutzen, P. J. 1970. The influence of nitrogen oxides on the atmospheric ozone content. *Quart. J. Roy. Met. Soc.* 96:320–25.

Crutzen, P. J. 1971. Ozone production rates in an oxygen-hydrogen-nitrogen oxide atmosphere. *J. Geophys. Res.* 76:7311–327.

Farman, J. C., B. G. Gardiner, and J. D. Shanklin. 1985. Large losses of total ozone in Antarctica reveal seasonal ClO_x/NO_x interaction. *Nature* 315:207–10.

Frederick, J. E. 2003. Ozone as a UV absorber. In *Encyclopedia of the Atmospheric Sciences,* eds. J. Holton, J. Pyle, and J. Curry, 1621–27. New York: Academic Press. 1621–27.

Goody, R. M., and J. C. G. Walker. 1972. *Atmospheres.* Englewood Cliffs, N.J.: Prentice–Hall.

Landsberg, H. 1981. *The Urban Climate.* New York: Academic Press.

Logan, J. 1985. Tropospheric ozone: Seasonal behavior, trends, and anthropogenic influence. *J. Geophys. Res.* 90:10, 463–482.

Molina, M. J., and F. S. Rowland. 1974. Stratospheric sink for chlorofluoromethanes: Chlorine atom-catalyzed destruction of ozone. *Nature* 249:810–12.

Seinfeld, J. H. and S. N. Pandis. 1998. *Atmospheric Chemistry and Physics.* New York: John Wiley and Sons.

Solomon, S. 1999. Stratospheric ozone depletion: A review of concepts and history. *Rev. Geophys.* 37:275–316.

U.S. Standard Atmosphere, 1976. Washington, D.C.: U. S. Government Printing Office, 1976.

Warneck, P. 1988. *Chemistry of the Natural Atmosphere.* New York: Academic Press.

World Meteorological Organization. 1991. *Scientific Assessment of Ozone Depletion: 1991.* Geneva: Global Ozone Research and Monitoring Project.

World Meteorological Organization. 2003. *Scientific Assessment of Ozone Depletion: 2002.* Geneva: Global Ozone Research and Monitoring Project.

Chapter 6

The Earth's Climate

INTRODUCTION

The concept of climate refers to the long-term statistics of all quantities required to describe the state of the atmosphere, including averages and indices of variability. Climate may apply to spatial scales from local to global and can encompass timescales ranging from decades to millions or even billions of years. The ability to define details of ancient climates degrades as one looks further into the past. The geologic record provides evidence of extended periods, spanning many millions of years, during which the planet remained relatively warm and essentially ice free, followed by similarly long periods when extensive ice sheets existed at high, and sometimes at middle, latitudes. These long-term climate shifts were linked to a slowly changing geographic distribution of land masses and the associated response of oceanic circulations.

During the last 1.8 million years, ice sheets have occasionally expanded from the polar region over the middle-latitude land masses of the Northern Hemisphere. The last such glacial advance reached its peak only 18,000 years ago, with the retreat occurring between 15,000 and 8000 years ago. Physical mechanisms associated with these relatively recent climatic changes include small variations in the solar constant, changes in the eccentricity of the Earth's orbit around the sun and the orientation of the rotation axis in space, changes in the planetary albedo, and variations in the magnitude of the greenhouse effect. Attempts to estimate the magnitude of global warming in response to increasing carbon dioxide abundances are complicated by the need to address the global cycling of water, possible changes in cloudiness and ice cover, and the cycling of carbon between the land, oceans, and living things.

6.1 A Definition of Climate

Previous chapters have considered the scientific principles necessary to describe the physical and chemical state of the Earth's atmosphere. The emphasis on fundamentals led to independent discussions of each process. For instance, the radiative equilibrium temperature profile derived in Section 2.7

took the abundance of all greenhouse gases as given. Yet, Chapter 3 showed that the pressure of water vapor, which is the most important longwave absorber, is a sensitive function of temperature. As another example, the geostrophic wind equations derived in Section 4.7 assume that horizontal pressure gradients are known. But spatial variations in pressure are related to the latitudinal dependence of radiative heating and cooling as well as to heat transport by the motions themselves. Clearly, the Earth's climate system involves the simultaneous, coupled action of all of the processes presented separately throughout this book.

How can one define the concept of climate? Strictly, climate involves all of the quantities required to describe the state of the atmosphere as introduced in Chapter 1. Consider a set of N different atmospheric properties denoted by $\{Q_i(\theta,\phi,t), i = 1,2,\ldots N\}$. In general, each property varies in latitude θ, longitude ϕ, and time t. Elements of this set include temperature, pressure, and relative humidity, among others. Strictly, the dependence of these quantities on altitude could be included as well, but for simplicity this discussion considers only conditions at the Earth's surface. Climate can be defined as the "long-term" statistics of the quantities that describe the state of the atmosphere (SMIC 1971). If a long-term average refers to a period of duration p, then the set $\{<Q_i(\theta,\phi,t,p)>, i = 1,2,\ldots N\}$ where

$$<Q_i(\theta,\phi,t,p)> = p^{-1} \int_t^{t+p} dt' \, Q_i(\theta,\phi,t') \qquad [6.1.1]$$

characterizes climate as a function of location (θ,ϕ) for the period extending from t to $t + p$. Based on this definition, climate can vary in time, although fluctuations on time scales shorter than p are averaged over. Variability during the averaging period could be defined by the standard deviation of the $Q_i(\theta,\phi,t')$.

Equation 6.1.1 allows climate to depend on location. It is possible to further average over limited ranges of latitude and longitude to develop climates for distinct geographic regions such as the tropics or the American Midwest. In the extreme one could envision a planetary scale climate by averaging over the entire Earth. The treatment of the radiative energy balance in Chapter 2 addressed this global scale and in doing so implicitly averaged over major variations in latitude and, to a lesser extent, in longitude.

A central point of the preceding discussion is that the concept of climate includes both a spatial scale and a temporal scale. Conditions in the atmosphere clearly vary over several spatial scales. The simple fact that the Earth is spherical leads to a systematic latitudinal variation in the amount of solar energy received by a unit horizontal area at the ground (Section 2.9). The result is a dependence of climate on latitude. Furthermore, the inhomogeneous nature of the Earth's surface leads to geographically diverse regional climates. The dimensions of land masses, bodies of water, ice sheets, and mountain ranges define natural spatial scales over which climatic variables can change.

The issue of temporal scales is particularly important. Two obvious timescales imposed on the atmosphere are those associated with the rotation of the Earth on its axis once every 24 hours and the orbital motion around the sun once every year. When combined with the tilt of the rotation axis relative to the plane of the orbit (Section 2.9), the latitude-dependent seasonal cycle results. If the timescale p used to define climate is very long compared to 1 year, the most familiar cycles of atmospheric variation average out. In modern terminology, applicable to the period of instrumental observations, an average over several decades or longer provides a reasonable operational definition of climate. With this definition and abundant meteorological data, it is possible to describe the climatology of a region, including the average state and the range of year-to-year variation. Indeed, indices of short-term variability can be included in the descriptions of climates over the last one or two centuries.

In considering climates of the distant past, it is essential to recognize the wide range of temporal scales used in their definition. As one probes further into history, to before the time of instrumental records, it is necessary to infer the characteristics of climate from information found in the geologic record. The ability to describe details of these earlier climates degrades rapidly. At best one can hope to estimate averages that may refer to centuries up to perhaps tens of millions or even hundreds of millions of years. In the framework of Eq. 6.1.1, the timescale p becomes very long, and the boundaries of the periods considered, t and $t + p$, become uncertain. Stanley (1986) and Pickering and Owen (1997) discussed how the geologic record can provide information from which to infer past environmental conditions on Earth. Regardless of the tools used to infer past climates, one conclusion is clear: the Earth's climate has undergone major changes over the planet's history, and these changes have occurred over a wide spectrum of timescales.

The following two sections provide a brief historical summary of Earth's past climates over widely different timescales. The first period encompasses the extended time from the formation of the planet approximately 4.6×10^9 years ago up to 1.8×10^6 years ago, an interval that encompasses roughly 99.96% of the Earth's history. The geologic information on climates during this time frame is incomplete at best and has poor temporal resolution. The second period, from 1.8×10^6 years ago to the present, encompasses the time during which species recognizable as ancestors of modern humans existed, but it represents less than 4/100 of 1% of the Earth's age. Information on climates during this time is of much higher quality than that for the earlier period.

6.2   Climate over Geological Timescales: 4.6 Billion to 1.8 Million Years Ago

The long span of geologic time called the "Precambrian" extends from the formation of the Earth to approximately 5.9×10^8 years ago. The Precambrian encompasses ancient volcanic activity that ejected gases from the planet's interior, the origin of the oceans and atmosphere, the formation and movement of land masses as several large tectonic plates that make up the Earth's outer crust moved about, the aging of the sun with consequent changes in the solar constant, and the rise of molecular oxygen to become a major atmospheric component. The ongoing evolution of the Earth's surface during the Precambrian destroyed much, but not all, of the geologic evidence needed to infer the nature of the Earth's earliest climates. Two facts, however, are apparent. First, before about 2.5×10^9 years ago, simple life in the form of bacteria and algae existed in the oceans. The temperature of the planet must have been sufficiently high for substantial amounts of water to remain in the liquid phase. Second, the global climate fluctuated over the Precambrian, becoming alternately cooler and warmer. Unfortunately, the sketchy available information does not provide a detailed understanding of the ancient changes in climate.

A thorough description of climate includes indices such as temperature, winds, and precipitation, but what geologic evidence of these quantities survives from the distant past? Changes in phase of water that occurred over continental scales provide such evidence. If a climatic cooling leads to the freezing of a large quantity of water on the planet's surface, the presence of geographically extended ice sheets can be detected long after the event has passed. The movement of a large volume of ice alters the underlying land. The advance of an ice sheet can move large boulders over long distances. Rocks beneath the ice will be crushed, and the retreat of the frozen mass leaves large indentations in the Earth's surface, which may become lakes. From this type of evidence it is possible to infer the existence of several prolonged cold periods called "ice ages" during which substantial amounts of ice existed on the Earth's surface, at least at high latitudes, and occasionally at middle latitudes as well. Relatively warm periods of long duration have occurred between the ice ages, during which the surface of the planet, including the polar regions, was ice free (SMIC 1971; Stanley 1986).

The ability to resolve events as a function of time is very limited in the Precambrian, and only coarse measures of the early climates exist. Subject to this limitation, the geologic record indicates that the Earth remained relatively warm from its earliest history to roughly 2.5 to 2.7×10^9 years ago. Extensive ice sheets then moved over the continents. The duration of this ancient glaciation is uncertain, but it surely persisted for tens of millions of years. It was followed by a period of order 1.5×10^9 years characterized by relative warmth and the absence of continental glaciers. Another period of extensive glaciation, which may be composed of several distinct episodes, occurred somewhere in the interval between 8×10^8 and 6×10^8 years ago. The geologic record points to the existence of ice over several continents during this time. The end of this ice age around 5.9×10^8 years ago marks the end of the Precambrian and the rapid evolution and spread of complex life forms over the planet.

The alternating cycles of temperature continued. The major ice age that ended with the Precambrian was followed by a prolonged interval of warmth, interrupted by a minor ice age between 4.6×10^8 and 4.3×10^8 years ago. The next major ice age, called the Karoo Ice Age, occurred somewhere in the period from about 3.5×10^8 to 2.5×10^8 years ago, and this glaciation persisted for 30 million to 50 million years. After the end of the Karoo Ice Age, paleotemperature data show an extended warm period. In the time between 2×10^8 and 6×10^7 years ago the annual surface temperature at both poles exceeded the melting point of ice (SMIC 1971).

The ice-free world of 60 million years ago was not to last, and a cooling began around 37 million years ago. Paleotemperature data reviewed by Stanley (1986) indicate a rapid drop in temperature, at least in some locations. This cooling is associated with the appearance and growth of ice on the Antarctic continent, which at that time occupied the south polar region, as it does today. For much of the time since the end of the Precambrian, the high-latitude regions of the planet have been free of ice (SMIC 1971). Thus, the very existence of polar ice can serve to define an ice age, irrespective of the latitudinal extent of this ice. Based on this definition, the Earth is still in an ice age that began approximately 37 million years ago.

Although the formation of ice began at high southern latitudes, creating the Antarctic ice sheet between 37 million and 24 million years ago, the temperature drop was experienced worldwide. The flow of energy in the oceans provides a mechanism for converting a local change in Antarctica into a global change. Ocean water in contact with Antarctic ice was cooled, became relatively dense, and sank. The cold water then spread north along the ocean bottom and eventually upwelled to the surface, thereby cooling geographic regions far removed from the south polar ice (Stanley 1986).

Climate change over the geological timescales considered in this section is linked to the evolution of the Earth itself. Over periods measured in tens of millions of years and longer, the geographic distribution of land masses changes in the process of continental drift. Major and prolonged ice ages of the distant past, such as that between 3.5×10^8 and 2.5×10^8 years ago, occurred when very large land masses, which today exist as separated continents, were merged together in the south polar region. The interior of the giant land mass was far removed from the warming influence of ocean currents and received little direct solar heating because of its high latitude. The consequence was widespread cooling and glaciation.

Changing ocean currents in response to the slow movement of land masses can also promote cooling. The motion of Australia away from Antarctica, to which it was linked before 60 million year ago, acted to disrupt the southward flow of water from temperate latitudes (Stanley 1986). This established a circumpolar flow that cut off much of the oceanic flux of heat that previously had warmed Antarctica. The onset of glaciation associated with the ice age that began 37 million years ago likely represents the combined effects of continental drift and changes in ocean circulation.

6.3 Climate over the Last 1.8 Million Years

Given the very long timescale of continental drift and accompanying changes in ocean circulation, these are not factors in driving climate change over the relatively brief period of the last 1.8 million years. Polar ice has persisted throughout this entire interval, defining it as an ongoing ice age. In addition, the period has been marked by the advance of glaciers to middle latitudes on several occasions followed by their eventual retreat.

Four or five major glacial periods and numerous lesser events have occurred during the last 1.8 million years. Their timing consists of characteristic superimposed periods of roughly 42,000 and 100,000 years, with changes of smaller magnitude at a period near 23,000 years (Hays et al. 1976). Furthermore, the time required for the glaciers to advance is much longer than the duration of their retreat. For example, the most recent glacial period began about 100,000 years ago. Glaciers over the Northern Hemisphere reached their most southerly extension around 18,000 years ago. The glacial retreat was relatively rapid and took place between 15,000 and 8,000 years ago.

At the time of maximum coverage during the most recent glacial period, ice existed over eastern North America down to a latitude of 38° or 39°, near the present-day location of St. Louis, Missouri, as well as over the Eurasian continent from the pole to approximately 50° north. The extensive buildup of continental ice sheets, measuring in excess of a kilometer thick, led to a lowering of the sea level by approximately 100 m, and the global average temperature was 8 to 10 K lower than now (SMIC 1971).

6.4 Mechanisms of Climate Change: An Overview

The previous sections made no attempt to provide a detailed historical account of climate change over the geological past; interested readers are referred to the comprehensive work by Stanley (1986) for this information. Rather, the goal here is to make the point that the Earth's climate has varied substantially over several timescales, including the period during which humans have inhabited the planet. The question to be addressed next is, What physical mechanisms might account for these past changes in climate, particularly those that occur on timescales of 100,000 years and shorter?

The causes of climate change may originate either external from or internal to the Earth-plus-atmosphere system. The obvious external factor is the amount of solar energy that crosses a unit horizontal area per unit time at the top of the atmosphere expressed as a function of latitude. The quantities involved here are the solar luminosity (Eq. 2.3.1), the distance between the sun and the Earth (Eq. 2.3.1), and the angle at which the solar beam intercepts a horizontal area (Fig. 2.13).

Factors involved in climate change that are internal to the Earth-plus-atmosphere system operate on a wide range of timescales. Continental drift, which produced long periods in the Earth's history when large land masses occupied the polar regions, was surely critical in causing the major ice ages over geologic timescales. Ocean currents accomplish a major transport of heat, and these circulations change as the positions of land masses gradually shift. Over the shorter timescale of centuries, oceanic motions respond to changes in water density associated with altered salinity, as can result from the melting of large expanses of ice.

At still shorter timescales, measured in decades, changes on the planet's surface or in the extent and thickness of cloud cover alter the albedo. The abundance and optical properties of particulate matter have an analogous influence via scattering and absorption. Finally, the abundances of gases that are radiatively active in the longwave part of the spectrum, such as CO_2 and H_2O, control the magnitude of the greenhouse warming experienced by the planet. A key issue here involves understanding the mechanisms that control the quantity of CO_2 that resides in the atmosphere during any given period. This encompasses modern concerns such as the burning of fossil fuels, but the exchange of CO_2 between the atmosphere and living things via respiration and photosynthesis, the biological processing of carbon into forms other than CO_2, and the storage of carbon in the oceans are all critical processes that must be treated.

A mechanistic understanding of past climate changes is complicated by the fact that all of these processes operate simultaneously, albeit over a wide range of timescales. Furthermore, it is not necessarily straightforward to distinguish cause and effect since various feedbacks exist. For example, an expansion of polar ice cover may alter the oceanic circulation, leading to a global cooling in climate. Simultaneously, the increase in albedo associated with the expanded ice cover reduces the amount of solar energy absorbed, thereby promoting further cooling and a still greater extension of the ice. The enhanced albedo is a cause of the ongoing expansion of the ice sheets, but the question of what caused the initial glaciation remains. This is where the actions of continental drift over geologic timescales and of changes in incoming solar energy over a variety of timescales become relevant.

6.5   Changes in Solar Luminosity and the Solar Constant

The solar constant S_E defined in Section 2.3 refers to a fixed Earth–sun distance so that temporal variations in S_E arise only from changes in solar luminosity (Eq. 2.3.1). As was the case with past changes in climate, the issue of timescales is central to the discussion. Over the last 4.6×10^9 years it is virtually certain that solar luminosity has increased substantially as the sun has aged, while over much shorter timescales the solar energy received at the top of the atmosphere for fixed solar distance can vary slightly over periods of decades.

Models of stellar evolution predict that solar luminosity should be increasing by several percent per billion years, giving a modern-day solar constant that is about 30% larger than that 4.6×10^9 years ago (Foukal 2004). The mechanism for this change over geologic timescales lies in nuclear fusion (Section 2.2). As the supply of hydrogen deep in the sun's interior gradually converts to helium in the process $4H \rightarrow He + $ gamma ray, the total solar number density decreases, leading to a contraction in size of the solar core. This contraction is accompanied by changes in temperature that appear as gradual increases in solar luminosity.

A solar constant equal to 70% of its current value, taken with the modern-day planetary albedo ($A = 0.3$) and greenhouse warming of about 33 K from Section 2.7, lead to an average surface temperature of only 266 K, which is below the freezing point of water. Based on this simple calculation it appears likely that the young Earth must have been subjected to added warming influences that are absent or weaker today. This warming may have come from a combination of a reduced albedo, a stronger greenhouse effect, and enhanced heat release from the Earth's interior.

Moving to vastly shorter timescales, data from satellite-based sensors show changes in the solar constant of very small amplitude associated with the well-known 11-year cycle in solar activity. Observations made from the late 1970s to the present reveal a good correlation between the solar constant and the number of sunspots visible on the solar disk. The incoming energy flux received by the Earth at "solar maximum," the time of largest sunspot numbers, is about 1.3 W m^{-2}, or 0.1%, greater than at "solar minimum" (Foukal et al. 2006). A historical record of sunspot numbers, albeit of varying quality, dates to the origin of the telescope in the early seventeenth century. This record can be used to estimate changes in the solar constant over the last 400 years by using statistical relationships derived from the recent era of satellite-based measurements. Although the 11-year sunspot cycle repeats backward through time, the amplitude of the cycle appears to change with an irregular period on the order of several decades to a century.

Data presented by Eddy (1977, 1981) show very large sunspot numbers during solar maxima, which occurred near 1778, 1870, 1947, and 1957. Unusually large sunspot numbers also characterized the solar maxima of 1979 and 1990. However, shortly after the start of observations in 1610, a very unusual period occurred between about 1640 and 1715, during which very few sunspots were visible. Today this period is called the "Maunder minimum," named after the scientist who focused attention on this anomaly. If the correlations between sunspot numbers and the solar constant observed today apply to the past, then the solar luminosity during much of the seventeenth century appears to have been slightly smaller than at present, although the difference would be measured in tenths of one percent. The Maunder minimum coincides with an unusually cool period, aptly named the Little Ice Age, whose existence is well documented in Europe (Eddy, 1981). Although it is of interest to suggest a link between the Maunder minimum and the Little Ice Age, it is uncertain whether their coincidence in time represents cause and effect or is accidental. The comprehensive review by Foukal et al. (2006) supports the latter conclusion. These authors found no evidence that variations in solar luminosity over the last 1000 years have been sufficient to cause a significant change in climate.

6.6   Changes in the Earth's Orbit and Orientation

For a fixed value of solar luminosity the amount of solar energy incident on a unit horizontal area in unit time at the top of the atmosphere depends on (1) the distance between the Earth and the sun and (2) the angle at which the incoming energy strikes the area. Regarding the first of these, Section 2.3 established that the Earth's orbit around the sun is slightly elliptical, and the present-day eccentricity is $\varepsilon = 0.0167$. Let the Earth–sun distance at any time be L (Eq. 2.3.1), where L_{MAX} and L_{MIN} are the maximum and minimum values of L, respectively, that exist during 1 year. The eccentricity is related to these extremes by $L_{MAX} = [(1 + \varepsilon)/(1 - \varepsilon)]L_{MIN}$, or, using the current value, $L_{MAX} = 1.034\,L_{MIN}$. Via Eq. 2.3.1 this gives an incoming solar flux at "perihelion," the time when $L = L_{MIN}$, which is 6.9% greater than at "aphelion," the time when $L = L_{MAX}$. Currently, perihelion occurs in early January and aphelion in early July. This annual cycle in incoming solar flux serves to modulate the amplitude of the seasonal cycle, which arises from the tilt of the Earth's rotation axis relative to the plane of the Earth's orbit around the sun.

The eccentricity of the Earth's orbit is not constant in time. Via Newton's Law of Universal Gravitation (Eq. 1.6.1), attractive forces exist between the Earth and all other bodies in the solar system, Jupiter, for example. Early in the twentieth century, the mathematician Milutin Milankovitch analyzed these gravitational forces and deduced both the magnitude and timescales associated with their effects on the Earth's orbit. Recent detailed calculations support the overall validity of Milankovitch's conclusions (Laskar et al. 2004). In addition, the recent results explicitly address limits on the ability to infer details of orbital parameters over timescales of tens of millions of years. These limits arise from interactions of gravitational forces exerted on the Earth by multiple extraterrestrial bodies.

An important conclusion of the gravitational analyses is as follows. Over a time of roughly 100,000 years, the eccentricity of the Earth's orbit varies from a value near $\varepsilon = 0$ to a maximum near $\varepsilon = 0.05$ and back to zero. When $\varepsilon = 0.05$, the incoming solar flux at perihelion is about 22% greater than at aphelion, although the average flux over a full orbital period is nearly independent of the eccentricity. This periodic eccentricity is the first contributor to "Milankovitch cycles" that are now believed to be significant factors in driving climate change over the last 1.8 million years (Broecker 1968; Hays et al. 1976; Pickering and Owen 1997).

Another effect of extraterrestrial gravitational forces concerns the orientation of the Earth's elliptical orbit about the sun. Imagine a line passing through the center of the sun that connects the point on the Earth's orbit where $L = L_{MIN}$ to the point where $L = L_{MAX}$. This line is the major axis of the Earth's orbit. The direction of the major axis defines the orientation of the orbit in space. An analysis of the gravitational forces that act on the Earth shows that the major axis rotates in space, with one complete rotation taking roughly 22,000 years. This rotation is about an axis hooked to the sun and oriented perpendicular to the plane of the orbit itself. The rotation of the orbit is one component of "precession." Note that precession has no direct connection to the specific location of the Earth along its orbit. The Earth completes one full orbit around the sun in a fixed time of 1 year irrespective of the orientation of this orbit.

What are the implications of the changing orbital orientation? Suppose that at an initial time $t_0(L_{MIN})$ the planet is at its closest approach to the sun, where $L = L_{MIN}$. At the later time $t_0(L_{MIN}) + 1$ year, the planet has completed one full orbit of the sun, no more and no less, but because of precession the distance of the Earth from the sun is not L_{MIN}. The time at which $L = L_{MIN}$ has moved to be slightly later than the time required to complete one orbit. Assume that the period of the orbit's precession is exactly 22,000 years, so the Earth will move around the sun 22,000 times as the elliptical orbit itself makes one full rotation in space. After 11,000 orbits of the sun, at the time $t_0(L_{MIN}) + 11,000$ years, the Earth–sun distance is L_{MAX}, while after 22,000 orbits, at time $t_0(L_{MIN}) + 22,000$ years, the distance has returned to L_{MIN}. As a result of precession, the specific date of the year on which the solar energy flux is a maximum moves through the year. Today this date is in early January; however, over the next 11,000 years it will gradually shift into early July, returning to January at the end of 22,000 years.

Another component of precession arises from a wobbling of the Earth's rotation axis. The Earth's axis does not point exactly in the same direction in space at all times. Instead, it rotates about a fixed direction like a spinning top, where the period, by coincidence, is near that associated with the rotation of the elliptical orbit. As a consequence of the wandering rotation axis, the time of, for example, summer solstice in the Northern Hemisphere (Fig. 2.14) does not occur at exactly the same point along the orbit, or date, from one year to the next. Instead the dates of the equinoxes and solstices shift gradually through the year, completing a full cycle in roughly the same time associated with the rotation of the elliptical orbit in space.

The changing orbital eccentricity and precession acting together lead to a periodicity in the incoming solar flux over 1 year. The amplitude of the annual cycle varies with a period of 100,000 years, and the date on which the maximum flux occurs slowly moves through the year with a period of approximately 22,000 years. The shifting dates of the equinoxes and solstices are imposed on these periodicities. The next orbital factor, which involves the tilt of the Earth's equator relative to the plane of the orbit, implies effects that are latitude dependent.

Section 2.9 makes the point that the seasonal cycle in temperature experienced at middle and high latitudes is a consequence of the nonzero angle between the plane of the Earth's equator and the plane of the orbit around the sun. This angle, called the "declination," is denoted here by δ, where the current value is $\delta = 23.4°$ to $23.5°$. During the timescale of a single orbit of the sun, the Earth's rotation axis points in the same direction in space (ignoring the small effect of precession identified earlier). This fact, combined with the nonzero declination, implies an annual cycle in the incoming solar energy flux on a horizontal surface located at any given latitude. Figures 2.13 and 2.14 depict the relevant geometry.

The annual cycle in solar flux on a horizontal surface is most pronounced at high latitudes. At latitudes from $90°$-δ to the Pole in either hemisphere, there will be periods when the sun is visible for a full 24-hour day, followed one half year later by 24-hour periods of darkness. At latitudes from δ south to δ north, the sun appears to pass directly overhead at noon twice per year, at one time moving toward the north and at the next moving toward the south. For values of declination typical of the Earth's, latitudes within this band will see little seasonal variation over a year. The magnitude of seasonal contrasts at middle and high latitudes clearly depends on the magnitude of δ. If $\delta = 0$, the seasonal cycle as known on the present-day Earth would vanish.

The final component of Milankovitch cycles arises because the declination δ varies with a period near 42,000 years. The minimum value of δ is slightly less than $22°$ and the maximum is near $24.5°$ (Broecker 1968). Values of δ near the small end of the range, which correspond to minimal seasonal contrasts, may be the most significant in promoting glacial advances from high to middle latitudes. Winters still will be sufficiently cold to allow liquid water to freeze, and the moderate temperatures associated with small δ values would allow larger atmospheric water vapor abundances to fuel precipitation to form the ice. During periods when δ is small, summers will be relatively cool and therefore more likely to allow ice formed during the winters to persist all year.

These described cycles, having periods near 100,000, 22,000, and 42,000 years, lead to predictable changes in the amount of solar energy received at any given latitude during one full orbit of the sun. Over the last 1.8 million years, as well as over longer geologic timescales, these cycles have undergone constructive and destructive interference with each other. This has resulted in times that promote glacial advances followed by warmer periods less favorable to glaciation at middle latitudes. That said, changes in the Earth's orbital parameters lead to only small changes in the solar energy flux that strikes a horizontal surface. Yet, the timing of glacial and interglacial periods shows a remarkable correlation to predictions based on the Milankovitch cycles (Hays et al. 1976). Today there is wide acceptance of a causal link between changes in the Earth's orientation and orbit on the one hand and changes in climate on the other. However, other variables must operate as well. For these, one must look to factors internal to the Earth-plus-atmosphere system.

6.7 Changes on Land and in the Atmosphere

The albedo A in Eq. 2.6.7 is an important determinant of the Earth's temperature. The global albedo, whose present value is $A = 0.3$, is an average over contributions from the planet's surface as well as from clouds and solid particles suspended in the atmosphere. The albedos of specific regions can range from less than 0.1 for open oceans when the sun is high in the sky (Budyko 1974), to 0.1 to 0.4 for various land surfaces (Liou 2002), to as large as 0.7 to 0.8 for snow and ice (Budyko 1974). Anthropogenic modification of the landscape can change the albedo in limited areas by, for example, replacing forested land with agricultural fields or urban development. Although changes in albedo from such activities may not be negligible, their strongest coupling to global climate likely comes through an influence on soil moisture and altered fluxes of heat and water vapor into the atmosphere (Pielke 2005).

Varying ice cover associated with glacial and interglacial periods surely has had a significant influence on the planet's albedo, and as noted in Section 6.4, changes in glacial extent constitute a "positive feedback" in the climate system. The high albedo of the ice reduces the amount of solar energy absorbed locally, and this promotes cooling and additional ice formation. Windblown dust, especially in combination with ice, provides another mechanism for altering the albedo of a region. Dust deposited on the surface of an ice sheet acts, through its smaller albedo, to promote heating from absorption of solar energy. This excess heating can initiate melting, which then accelerates via the feedback identified earlier.

Trace gases play a major role in the Earth's climate system. The radiative equilibrium temperature derivation in Section 2.7 assumed the abundances of all gases that absorb terrestrial longwave radiation to be known. An important molecule in this regard is carbon dioxide. What processes determine the atmospheric abundance of this gas? The reasoning required here is similar to that presented in Chapter 5, which considered the chemical production and loss of atmospheric gases. In the case of CO_2 the source and sink are expressed as upward and downward fluxes, respectively, summed over the Earth's surface, plus a small production by in situ air chemistry. The exchange of atmospheric CO_2 with the surface is one part of a larger global carbon cycle that involves the deep oceans, sediments, and the solid Earth. Carbon is stored in these reservoirs in forms other than CO_2. For this reason, it is convenient to measure fluxes in terms of the mass of carbon atoms per year, g C year^{-1}, exchanged between these reservoirs as opposed to the mass of CO_2. The following summary focuses on those pieces of the carbon cycle that pertain directly to the atmosphere.

The processes of photosynthesis and respiration appeared in Section 1.2 in the context of their influence on the abundance of atmospheric oxygen, but they also have a major role in partitioning carbon between the atmosphere, where the atom resides as CO_2, and the terrestrial biosphere, where it exists as organic matter. Values of fluxes estimated in the literature vary considerably (e.g., Seinfeld and Pandis 1998; Walther 2005), but there is agreement that the globally averaged downward flux of atmospheric CO_2 to the land is slightly larger than that in the opposite direction (by perhaps 1.0 to 2.5%) and that both fluxes are of the order of 1×10^{17} g C year^{-1}. Photosynthesis, which draws CO_2 out of the atmosphere, is responsible for the downward flux, whereas respiration, decay of organic matter, and the burning of biomass provides a source at the ground that supplies the upward flow. Figure 6.1 depicts these, and other, components of the carbon cycle.

The atmosphere and oceans also exchange CO_2. A fraction of the CO_2 molecules that strike the ocean surface dissolve in the liquid, thereby providing a downward flux from the air into the water. Aquatic organisms then convert a portion of the carbon into organic forms as well as into calcium carbonate, contained in seashells. Finally, as dissolved CO_2 reaches the water–air interface, some of the molecules break free of the liquid phase and produce an upward flux into the atmosphere. These fluxes of CO_2 between

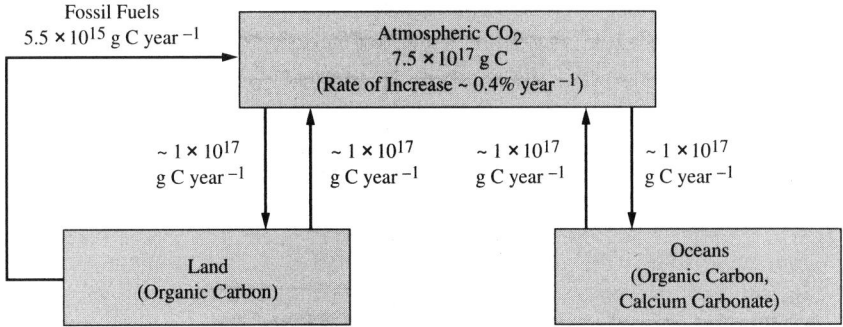

FIGURE 6.1   The exchange of carbon dioxide between the atmosphere, land, and oceans. Natural fluxes between the land and atmosphere and between the oceans and atmosphere are nearly in balance. The burning of fossil fuels creates an additional flux from land to atmosphere and an upward trend of 0.4% year^{-1} in atmospheric CO_2 abundance.

the atmosphere and ocean are also of the order 1×10^{17} g C year^{-1}, with the downward flux again slightly larger than the upward by about 2%.

Based solely on the estimates given thus far, the downward flux of CO_2 into the land and oceans exceeds that moving upward into the atmosphere. The values given by Seinfeld and Pandis (1998) are consistent with a net loss of CO_2 from the atmosphere near 2.3×10^{15} g C year^{-1}. The significant missing piece of the atmospheric CO_2 budget involves the burning of fossil fuels, which provides an additional upward flux at the ground. Section 5.10 shows that an efficient combustion process releases CO_2 into the atmosphere, whereas a less efficient process releases CO. However, once in the atmosphere, CO converts into CO_2 in R-30. The result is a surface-to-air flux equivalent to approximately 5.5×10^{15} g C year^{-1} (Seinfeld and Pandis 1998; Walther 2005). This leads to a net increase in the atmospheric CO_2 abundance of 3.2×10^{15} g C year^{-1}, or about 0.4% per year of the entire atmospheric reservoir of 7.5×10^{17} g C.

It is virtually certain that the natural fluxes of carbon between the atmosphere, ocean, and land have varied over all of the timescales considered in this chapter. The abundances of plant and animal life that perform photosynthesis and respiration, respectively, have varied over geologic timescales on both the land and in the oceans. Furthermore, the amount of CO_2 that can dissolve in the oceans is temperature dependent, becoming greater as the waters cool. Finally, the efficiency with which carbon is converted into organic forms and calcium carbonate depends on the availability of various nutrients that enter biological processing and on temperature.

Given the complexity of the biogeochemical cycling of carbon, it is not surprising that atmospheric CO_2 amounts have fluctuated over time. Interestingly, glacial periods appear to be extended times of relatively low atmospheric CO_2 abundances, whereas interglacial periods are the opposite (Petit et al. 1999). Whether the accompanying changes in the strength of the greenhouse effect are a primary cause of variations in temperature or act as a reinforcing effect of preexisting glacial advances is unclear. Over the much shorter timescale since the start of the Industrial Revolution, it is virtually certain that the observed increase in atmospheric CO_2 mixing ratio, from near 280 ppmv (Houghton et al. 1996) to 375 ppmv, reflects an imbalance in the cycling of carbon caused by the burning of fossil fuels.

Human activities have had a direct effect on the atmospheric abundance of CO_2, and this is the origin of current concerns about global warming. Still, water vapor is the most important greenhouse gas in the atmosphere. Chapter 3 addressed the processes that control the atmospheric water vapor abundance. To see how changes in carbon dioxide and water can act together to alter climate, consider Eqs. 2.6.7 and 2.7.12. The combination of these two expressions provides an estimate of the Earth's globally averaged surface temperature T_G:

$$T_G = [\,(1 + \tau)\,(1 - A)\,S_E\,/(4\sigma)\,]^{1/4} \qquad [6.7.1]$$

The temperature varies with the solar constant (S_E), the albedo of the Earth-plus-atmosphere system (A) and the longwave extinction thickness (τ), which includes effects of all greenhouse gases and clouds. Although highly simplified, this model allows one to identify mechanisms by which human activity might lead to changes in the Earth's temperature. The burning of fossil fuels is increasing the CO_2 content of the atmosphere, so τ is growing over time. Based solely on the CO_2 abundance, the effect on temperature is straightforward. As τ increases, T_G must increase as well. However, in practice there are several feedback mechanisms that make it difficult to predict how the Earth's surface temperature will change in response to increasing CO_2 amounts, and these involve the properties of water.

Figure 6.2 illustrates some of these feedbacks. The initial change involves release of CO_2 into the atmosphere from the burning of fossil fuels. Figure 6.2 depicts this by $\Delta CO_2 > 0$. The increase in CO_2 abundance increases the longwave extinction thickness, where the relationship between a change in the CO_2 amount and a change in τ can be calculated from detailed knowledge of the molecule's radiative properties. Qualitatively, one can write $\Delta\tau > 0$. Then the model of the greenhouse effect shows that an increase in τ leads to an increase in surface temperature, $\Delta T_G > 0$. This is where the simple model stops and the complications begin. First, oceans cover most of

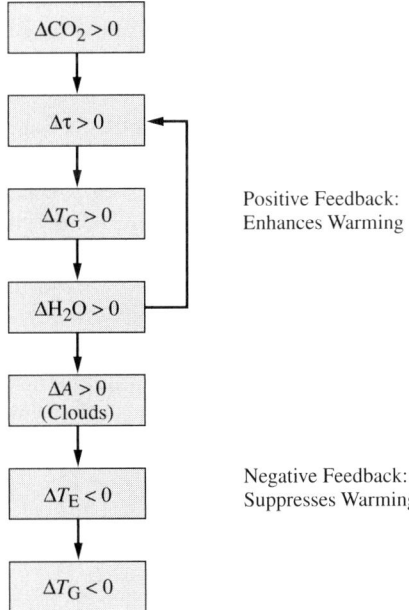

FIGURE 6.2   Climate feedback mechanisms. An increase in the atmospheric CO_2 abundance causes an increase in surface temperature. This change alters the efficiencies of other processes such as evaporation that influence temperature. The feedback may increase or decrease the initial warming.

the Earth. An increase in the temperature of the ocean's surface increases the rate at which water evaporates into the air. As a first estimate, one expects a warmer atmosphere to contain more water vapor, and H_2O is a very efficient greenhouse gas. An increased water vapor amount, $\Delta H_2O > 0$, increases the longwave extinction thickness in a way similar to increasing CO_2. To treat this feedback, it is necessary to model all of the processes that control the quantity of water vapor in the atmosphere. This means understanding all components of the global water cycle identified in Section 3.10, including evaporation, cloud formation, and precipitation. Most of the global warming computed by current models comes from the increased abundance of atmospheric water vapor and not from CO_2. The rise in CO_2 abundance from fossil fuel burning initiates the warming, but in the end most of the temperature increase is associated with H_2O. As with changes in ice albedo discussed in Section 6.4, this is a positive feedback in which an initial warming triggers changes that cause still more warming.

There are also negative feedbacks in the climate system, and these act to suppress the magnitude of a warming. Water vapor is at the center of one of these as well. If, on a warmer world, more water vapor evaporates and enters the atmosphere, there is a possibility that cloudiness will change, although even the sign of this response is uncertain. If, for example, clouds become thicker or cover a larger fraction of the planet, the Earth's albedo will increase. Figure 6.2 depicts this as $\Delta A > 0$, and this change leads to a lowering of the surface temperature, $\Delta T_G < 0$. To further complicate the situation, an increase in cloudiness would increase the longwave extinction thickness and produce a positive feedback that promotes warming. Should cloudiness decrease, the opposite sequence of events takes place. Based on qualitative reasoning alone, it is not possible to tell whether the positive feedbacks or the negative feedbacks are more important. A detailed treatment of cloud properties such as water content, the size distribution of droplets, and temperature is necessary to determine the magnitude and sign of this effect.

The best quantitative models available indicate that increasing atmospheric CO_2 amounts will lead to a net global warming (Houghton et al. 1996; Archer 2006), although predictions are complicated by the noted water vapor feedbacks, as well as by changes in heat storage in the oceans, ice cover, energy transport by winds and ocean currents, and even by potential changes in CO_2 uptake in photosynthesis. For all of these reasons, quantitative estimates of future climate change in response to anthropogenic and natural causes are subject to significant uncertainties.

6.8   References

Archer, D. 2006. *Global Warming: Understanding the Forecast.* Oxford: Blackwell Publishing Ltd.

Broecker, W. S. 1968. In defense of the astronomical theory of glaciation. In *Causes of Climate Change,* ed. J. M. Mitchell, Jr., 139–41. Boston: American Meteorological Society.

Budyko, M. I. 1974. *Climate and Life.* New York: Academic Press.

Eddy, J. A. 1977. Historical evidence for the existence of the solar cycle. In *The Solar Output and its Variation,* ed. O. R. White, 51–71. Boulder: Colorado Associated University Press.

Eddy, J. A. 1981. The search for solar history. In *Fire of Life,* 108–15. New York: W. W. Norton and Co.

Foukal, P. V. 2004. *Solar Astrophysics.* 2nd ed. New York: Wiley-VCH.

Foukal, P., C. Frohlich, H. Spruit, and T. M. L. Wigley. 2006. Variations in solar luminosity and their effect on Earth's climate. *Nature* 443:161–6.

Hays, J. D., J. Imbrie, and N. J. Shackleton. 1976. Variations in Earth's orbit: Pacemaker of the ice ages. *Science* 194:1121–32.

Houghton, J. T., L. G. Meira Filho, B. A. Callandar, N. Harris, A. Kattenberg, and K. Maskell, eds. 1996. *Climate Change 1995—The Science of Climate Change.* Cambridge, UK: Cambridge University Press.

Laskar, J., P. Robutel, F. Joutel, M. Gastineau, A. C. M. Correia, and B. Levrard. 2004. A long-term numerical solution for the insolation quantities of the Earth. *Astronomy and Astrophysics* 428:261–85.

Liou, K. N. 2002. *An Introduction to Atmospheric Radiation.* New York: Academic Press.

Petit, J. R., J. Jouzel, D. Raynaud, N. I. Barkov, J.-M. Barnola, I. Basile, M. Benders, J. Chappellaz, M. Davis, G. Delayque, M. Delmotte, V. M. Kotlyakov, M. Legrand, V. Y. Lipenkov, C. Lorius, L. Pépin, C. Ritz, E. Saltzman, and M. Stievenard. 1999. Climate and atmospheric history of the past 420,000 years from the Vostok ice core, Antarctica. *Nature* 399:429–36.

Pickering, K. Y., and L. A. Owen. 1997. *An Introduction to Global Environmental Issues.* 2nd ed. New York: Routledge.

Pielke, R. A., Sr. 2005. Land use and climate change. *Science* 310:1625–26.

Seinfeld, J. H., and S. N. Pandis. 1998. *Atmospheric Chemistry and Physics.* New York: John Wiley and Sons.

SMIC. 1971. *Inadvertent Climate Modification.* Cambridge, Mass.: The MIT Press.

Stanley, S. M. 1986. *Earth and Life Through Time.* New York: W. Freeman and Co.

Walther, J. V. 2005. *Essentials of Geochemistry.* Sudbury, Mass.: Jones and Bartlett.

Index